NUCLEAR MAGNETIC RESONANCE IN FERRO- AND ANTIFERROMAGNETS

NUCLEAR MAGNETIC RESONANCE IN FERRO- AND ANTIFERROMAGNETS

by
E.A. Turov and M.P. Petrov

Translated by E. HARNIK
The Hebrew University of Jerusalem

ISRAEL PROGRAM FOR SCIENTIFIC TRANSLATIONS
JERUSALEM · LONDON

HALSTED PRESS
A DIVISION OF
JOHN WILEY & SONS, INC.
NEW YORK

Sole distributors for the Western Hemisphere

HALSTED PRESS, a division of
JOHN WILEY & SONS, Inc., NEW YORK

Distributed in the rest of the world by

KETER PUBLISHING HOUSE, JERUSALEM
KETER PUBLISHERS, LONDON

Library of Congress Catalog Card Number 72–4099
ISBN   0 7065 1105 0
IPST Catalog Number 2314

This book is a translation from Russian of
YADERNYI MAGNITNYI REZONANS V FERRO- I ANTIFERROMAGNETIKAKH
Izdatel'stvo "Nauka"
Moscow, 1969

Preface

Although there are many good books on nuclear magnetic resonance (NMR), none of them actually considers the problem of NMR in ferro- and antiferromagnets. Meanwhile investigations in the last six–seven years have shown that NMR exhibits striking peculiarities in these magnetic crystals, and that it appears to be one of the most promising methods for investigating the nature of the magnetically ordered state of matter.

Apparently, experimental data obtained with the use of NMR and corresponding theoretical problems will in the future be included in all textbooks on ferro- and antiferromagnetism. In the meantime, however, they have not been discussed even in comprehensive monographs.

Convinced of the necessity to introduce NMR methods into the physics of magnetic materials, the authors wrote this book with a double aim: on the one hand, to raise the interest of NMR specialists in the fascinating physical problems which exist in this field, and, on the other, to bring to the attention of the physicist working on magnetism the great possibilities revealed by the NMR methods. Unfortunately, there is a communication gap between the representatives of these two groups of investigators; they usually work in different scientific institutions and they are often little acquainted with the theoretical ideas and with the research projects of other groups.

The foregoing constituted the main difficulty which the authors have faced while writing this book. We wanted to write just a small book devoted to a restricted problem, a book that could be read through in a few days, without forgetting at the end what had been said at the beginning. At the same time, the double address of the book (to specialists in NMR and in ferromagnetism) seemed to require, on the one hand, a description of the basic ideas and methods of the theory of ferro- and

antiferromagnetism (chiefly the theory of spin waves) and, on the other, the clarification of many concepts in the NMR theory. All these threatened to increase the volume of the book, and the authors doubt very much whether they have succeeded in finding the "golden mean." The authors will be satisfied if this book partially fills the existing gap between textbooks on NMR and those on ferromagnetism (but in no way replaces them).

Many readers may wish to get acquainted with the features and applications of NMR in ferro- and antiferromagnets only in a general way, without going deeply into details. They need to read only the first chapter, which is in fact a concise theoretical review of all the general problems (the experimental physicist should also read Chapter X, which reviews the various features of the NMR apparatus used with these materials).

The list of references is by no means intended to be complete. The references were chosen according to the following principles. Works which advanced some new physical (or methodological) ideas, directly related to the subject of the book, were the first to be chosen. Then came articles which supplied experimental data either used as examples in the discussion of theoretical laws and estimates in text, or listed in the table of resonance frequencies and local fields in Appendix II. It frequently happens that out of a number of publications of the same (or sometimes different) authors, only the latest work is mentioned, provided it contains, in the main, all previous information. Finally, as regards references to books, reviews and original papers on the auxiliary problems of solid state physics and NMR, preference was given to sources most accessible to the reader (good presentation, new edition, etc.). Of course, the principles mentioned do not apply to the works of the present authors; indeed, it is difficult not to quote oneself.

Chapters I–VII and Chapter IX were written by Turov, and Chapters VIII and X by Petrov.

E. Turov, M. Petrov

Table of Contents

CHAPTER I

Introduction. Specific Features of NMR in Ferro- and Antiferromagnets

1. Local Magnetic Fields at the Nuclei

The development of the physics of magnetically ordered media (ferromagnets, antiferromagnets, etc.) requires today, more than ever, the application of such methods of investigation that would supply information on the *local spatial* distribution of electron and spin densities in the solid state. Such local characteristics can be derived through the study of nuclear magnetic resonance (NMR). Of course, there are the more direct methods of diffraction (neutron or X-ray). NMR, however, has some advantage over these, associated with the great accuracy inherent in r.f. spectroscopy.

A nucleus possessing a magnetic moment* constitutes a natural point probe, placed in the electron system of the crystal, by means of which much information on this system can be obtained with the use of NMR methods. Speaking of NMR in ferro- and antiferromagnets, what one usually has in mind is first of all the measurement by means of NMR of the local magnetic fields set up by the electrons at the nuclei. The Mössbauer effect is employed for this same purpose (e.g., Collection, 1962; Freeman and Watson, 1965). The nuclei of many magnetic atoms which constitute the main interest of physicists working on magnetism do not, however, possess Mössbauer levels.

---

* It should be remembered that this includes all odd nuclei (i.e., nuclei with odd mass numbers $A = N + Z$), and also the so-called odd-odd nuclei H^2, Li^6, B^{10} and N^{14}, for which both the number of neutrons (N) and the number of protons (Z) are odd. The list of basic magnetic nuclei on which NMR experiments were carried out (or should be carried out) is presented in Appendix I.

1

An electron, situated at point r and having orbital momentum l and spin s, creates at the nucleus (at the point $r = 0$) a local magnetic field described by the operator (Abragam, 1963)

$$H_e = -g\mu_B \left\{ \frac{l-s}{r^3} + 3\frac{r(rs)}{r^5} + \frac{8\pi}{3}s\delta(r) \right\}, \tag{1.1}$$

where μ_B is the Bohr magneton and g is the spectroscopic splitting factor of the electron. The term with the delta function $\delta(r)$ is called the Fermi contact field, and the other two terms constitute the fields created by a point dipole $g\mu_B s$ and by the orbital current of electron with angular momentum l.

The calculation of the expectation value of the field $H_{loc} = \sum_i \int \psi^* H_e^i \psi d\tau$ of all the electrons requires knowledge of the distribution of electron and spin densities in the atom (i.e., its wave function ψ) and also, in general, in the crystal. Such calculations are carried out by the Hartree–Fock method with the use of computers. There exist a number of reviews in literature devoted to the nature of local magnetic fields at the nuclei (e.g., Watson and Freeman (1961, 1967) and Freeman and Watson (1965)). Accordingly, we shall confine ourselves here only to the enumeration of the basic conclusions which are summarized in these reviews.

In solids the local field at the nuclei of transition elements of the iron group, with unpaired spins in the 3d-shell ("3d-elements"), is basically produced by the Fermi contact interaction. The latter vanishes for all except the s-electrons, since the electron density at the nucleus $\rho(0) \equiv |\psi_s(0)|^2 \neq 0$ for them only. However, to obtain a nonzero value for H_{loc}, the spin density of the s-electrons must also be nonzero at the nucleus, i.e.,

$$|\psi_{s\uparrow}(0)|^2 - |\psi_{s\downarrow}(0)|^2 \neq 0.$$

Calculation showed that in the "3d-elements" the main contribution to the local field came from electrons of the inner s-shells spin-polarized by their exchange interaction with the electrons of the uncompensated 3d-shell. In metals and alloys the exchange interaction between the 3d-shell electrons and the conduction electrons ("4s-electrons") leads also to the spin polarization of the latter. This mechanism introduces a contribution into the Fermi contact field the sign of which is opposite to that of the inner s-electron contribution. The magnitude of H_{loc} at the nuclei of "3d-elements" attains a value of 10^5–10^6 Oe.

In the case of rare earth elements, the uncompensated 4f-shell which has a considerably smaller radius than the 3d-shell of the iron group elements (and an unlocked orbital momentum), the main contribution to H_{loc} comes from the orbital and dipole field of the 4f-electrons at the nucleus. Here the field at the nucleus can attain a value of 10^6–10^7 Oe.

The metallic bond together with the covalent effects of transition metals produce

considerable local fields (10^4–10^5 Oe) also at the nuclei of nonmagnetic atoms in magnetic alloys and intermetallic compounds.

Similarly, in chemical compounds in which paramagnetic ions participate, the chemical bond and, in particular, effects of covalency and overlap of electron shells lead to the spin polarization of the electron shells of the nonmagnetic ions. This may be one of the reasons for the appearance of local fields at the nuclei of nonmagnetic ions in such compounds (of the order of 10^4 Oe).

All the local fields enumerated above which are associated with the interaction of the nuclear spin with its "own" electron shell (including also polarization effects caused by neighboring magnetic ions) will be called hyperfine fields and the corresponding interaction—hyperfine interaction (HFI). It is necessary to distinguish them from magnetic fields caused by *direct* interaction between nuclear spins and electron shells of "foreign" magnetic atoms. The main source of such fields is often the dipole–dipole interaction between the nuclear magnetic moment and the total momenta of the surrounding magnetic ions. These fields will henceforth be referred to as dipole fields. The magnitude of the latter does not exceed a few kilooersted.

The interaction between the nuclear spin I of a magnetic atom and the electrons of an uncompensated $3d$- (or $4f$-) shell can be represented by the HFI Hamiltonian

$$\mathscr{H}_{HFI} = I \cdot AS, \tag{1.2}$$

where A is the HFI constant and S is the total spin of the atom. In the case of rare-earth elements the total angular momentum of the atom $J = S + L$ replaces S. It should be pointed out that when the nuclei in a crystal are situated at lattice sites the surroundings of which do not possess cubic symmetry, the parameter A is in general a tensor having the corresponding symmetry. However, this anisotropy of the HFI is usually small.

The HFI for nuclei of nonmagnetic ions, caused, in particular, by the polarization of the filled shells can, by analogy with (1.2), be represented in the form

$$\mathscr{H}_{HFI} = I \cdot \sum_j A(R_j) S_j, \tag{1.2a}$$

since the above mentioned polarization is proportional to the spin of the magnetic ions. The summation in (1.2a) is carried out over the lattice sites of the nearest neighbor magnetic ions.

For the sake of completeness we shall also write down the Hamiltonian of the dipole interaction between the nuclear magnetic moment $\gamma_n \hbar I$ and the total momenta $\gamma_e \hbar J_j$ of the surrounding magnetic atoms or ions:

$$\mathscr{H}_{dip} = -\mu_n I \cdot \mu_e \sum_j \left(\frac{J_j}{R_j^3} - 3 \cdot \frac{R_j(J_j R_j)}{R_j^5} \right), \tag{1.3}$$

where $\mu_n = \gamma_n \hbar$ and $\mu_e = \gamma_e \hbar$, while γ_n and γ_e are the gyromagnetic ratios for the nucleus and the electron, respectively ($\gamma_e < 0$, but γ_n is positive in the majority of cases. See Appendix I).

The concept of the magnetic sublattice is usually used in the description of the magnetically ordered state of a crystal (in ferri- and antiferromagnets). A magnetic sublattice is made up of all those magnetic ions the position and magnetic moments of which are related by translational symmetry with the period of the magnetic unit cell. In other words, the magnetic sublattice comprises all magnetic ions which are equivalent with respect to their crystallochemical as well as magnetic properties. In many cases the magnetic structure of a crystal can be represented as an aggregate of a few interpenetrating magnetic sublattices. If the magnetization $M(r)$ of each magnetic sublattice (defined as the local density of its magnetic moment) is introduced separately, the density of the HFI energy of one sublattice can be written in the form

$$\mathscr{H}_{\mathrm{HFI}}(r) = A_0 M m, \qquad (1.4)$$

where $A_0 = A/(N\gamma_e\gamma_n\hbar^2)$, N is the number per unit volume of the magnetic ions of the given sublattice and m is the nuclear magnetization for the same sublattice.*

Now the hyperfine part of the local field, acting on the nucleus, has the form

$$H_{\mathrm{loc}} = -\frac{\partial \mathscr{H}_{\mathrm{HFI}}}{\partial m} = -A_0 M.$$

Let us emphasize that this expression is applicable to the case of simple ferromagnets (with one magnetic sublattice) as well as to the case of ferri- and antiferromagnets, if M is taken to denote the magnetization of the corresponding sublattice.

Experiment shows that in the majority of cases the field at the nuclei H_{loc} opposes the local magnetization. This means that in such cases $A_0 > 0$ (and, consequently, $A < 0$).

As has already been shown, the instantaneous magnitude of H_{loc} at the nuclei of magnetic atoms attains the value of 10^5–10^6 Oe. However, the thermal motion produces fluctuations in the directions of the magnetic moments while the exchange and other interactions between them lead to an averaging of the moment of each individual atom. As a result, the nuclear spin "sees" at a given temperature some average value of H_{loc}:

$$\langle H_{\mathrm{loc}} \rangle = -A_0 \langle M \rangle. \qquad (1.5)$$

---

* The relation (1.4) can be considered as a special case of a more general expression

$$\mathscr{H}_{\mathrm{HFI}} = \int \mathscr{H}_{\mathrm{HFI}}(r)\,dr = \int A_0(r - r')\,M(r)\,m(r')\,dr\,dr', \qquad (1.4a)$$

which appears to be analogous to relation (1.2a) in terms of a continuous medium.

The NMR frequency ω_n is computed as the Larmor precession frequency in the resultant static magnetic field acting on the nuclear magnetic moment:

$$\omega_n = \gamma_n \left| H - A_0 \langle M \rangle \right|, \tag{1.6}$$

where H is the external magnetic field.*

In the paramagnetic phase $\langle M \rangle = \chi_p H$, where χ_p is the paramagnetic susceptibility of the electron system. Consequently, the NMR frequency in the paramagnet is

$$\omega_n = \gamma_n (1 - A_0 \chi_p) H,$$

and the local fields at the nuclei induced by the HFI lead to only a relatively small shift in the resonance frequency:

$$\delta \omega_n / \omega_n = A_0 \chi_p \ll 1,$$

analogous to the Knight shift.**

At sufficiently low temperatures in ferro- and antiferromagnets the absolute value of the average local magnetization is $|\langle M \rangle| \simeq M_0$, where M_0 is the magnetization of the corresponding sublattice at the temperature $T \to 0°K$ (in the presence of domain structure the direction of $\langle M \rangle$ can be considerably nonuniform in the bulk of the specimen). Therefore $|\langle H_{\text{loc}} \rangle| \simeq A_0 M_0$ will be close to its instantaneous value (i.e., to the quantum-mechanical expectation value), at low temperatures in any case.

2. Basic Features of NMR in Magnetically Ordered Crystals

The presence of a large local field at the nuclei of magnetic (and in a number of cases also of nonmagnetic) ions in ferro- and antiferromagnets leads to a series of peculiarities in their NMR phenomena in comparison with other crystals. The most obvious and simple of these are the following two.

The first is connected with the static (longitudinal) component of the local field

$$H_n \equiv \langle H_{\text{loc}} \rangle_Z = - A_0 \langle M_Z \rangle \tag{1.7}$$

(Z is the direction of the quantization axis of the electron moment at the site of the nucleus of interest). The NMR frequency in this field

$$\omega_{n0} = |\gamma_n H_n|$$

---

* Actually H includes also the dipole fields, but for the sake of simplicity this will not be taken into consideration here (see Chapter II, Sec. 1).

** It should be noted, however, that this shift is considerably larger than the Knight shift in metals or the chemical shift in diamagnetic materials. Besides, it must be born in mind that in general A_0 and χ_p are tensors.

is two–three orders of magnitude larger than in the paramagnetic phase of the same crystals in typical magnetic fields of 10^3–10^4 Oe. Consequently, here, in contrast to paramagnets, the main contribution to the resonance frequency (1.6) comes from the second term, while the external magnetic field will only introduce an insignificant shift in the NMR line. It should be borne in mind that in the case of ferrimagnets (or antiferromagnets) the sublattices of which possess oppositely directed magnetic moments, the field H will add to H_n in one sublattice but will subtract from it in the other (if, of course, H is directed along the axis of magnetic ordering). In multi-domain samples when the external magnetic field is compensated by the demagnetizing (dipole) fields, the effect of H on the NMR frequency will be even smaller.

The problem of NMR frequency and the influence of different factors on it will be described in more detail in Chapter II.

The other effect is connected with the dynamic (transverse) component of H_{loc}. The quantum resonant transitions of the nuclear spins are mainly induced not by the direct action of the *external* variable magnetic field h_x (with a frequency $\omega \sim \omega_n$) but by the variable part of the internal local field H_{loc} set up by the field h_x.

As a matter of fact, the variable (transverse) component of H_{loc} is given according to (1.5) by

$$H_{loc}^{\perp} = -A_0 M_{\perp}.$$

Let us again consider, for instance, a single-domain ferromagnetic specimen magnetized to saturation along the easy axis.

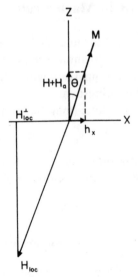

Figure 1

Illustrating the determination of the amplification coefficient of the radiofrequency field $\eta = H_{loc}^{\perp}/h_x$.

Then (Figure 1)

$$M_\perp \simeq M_0\Theta \simeq M_0 \frac{h_x}{H + H_a} = \chi_{rot}h_x,$$

where $\chi_{rot} = M_0/(H + H_a)$ is the magnetic susceptibility with respect to the field $h_x \ll (H + H_a)$, associated with the rotation of the magnetization under the action of this field, and H_a is the effective field of magnetic anisotropy.*

Hence, in addition to h_x, the following variable field acts on the nuclear spins

$$H_{loc}^\perp \simeq H_n\Theta = \frac{H_n}{H + H_a} h_x = - A_0\chi_{rot}h_x = \eta h_x. \tag{1.8}$$

The magnitude

$$\eta = - A_0\chi_{rot} = \frac{H_n}{H + H_a} \tag{1.9}$$

is called the amplification coefficient of NMR in a single-domain specimen. Assuming $H_n \sim 10^5{-}10^6$ and $(H + H_a) \sim 10^3{-}10^4$ Oe we have $\eta = 10{-}10^3$.

In a domain wall the nucleus "sees" even a larger variable field. The fact is that the magnetization of a multidomain ferromagnet in a weak variable magnetic field takes place via the reversible displacement of the domain walls. Through the displacements the local magnetization inside the domain wall experiences a particularly strong rotation; the angle of rotation (to which $H_{loc}^\perp$ is proportional) is maximum in the center of the wall (Figure 2). As a matter of fact, it turns out that, as regards its order of magnitude, the effective amplification coefficient η_{eff} for a

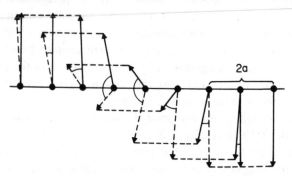

Figure 2

Diagram of a 180° domain wall. A displacement of the wall by two interatomic distances is shown.

---

* It is well known that the energy of a ferromagnetic crystal depends on the direction of the spontaneous magnetization with respect to the crystallographic axes. The direction for which this energy is minimal is called the easy axis. The existence of an easy axis can most simply be taken into consideration by the introduction of an effective field of magnetic anisotropy H_a, directed along the corresponding easy axis. To be more precise, the direction of H_a is in every case chosen in such a way that the energy of the magnetic moment in this field (the energy of magnetic anisotropy) $E_a = - MH_a$ is a minimum.

nucleus in the domain wall differs from (1.9) in the replacement of the "rotation susceptibility" χ_{rot} by the initial (reversible) displacement susceptibility χ_{dis} (Portis and Gossard, 1960)

$$\eta_{eff} \sim - A_0 \chi_{dis}.$$

Since χ_{dis} is one to two orders of magnitude larger than χ_{rot}, this means that the alternating field $H_{loc}^\perp = \eta_{eff} h_x$, which acts on the nuclear spins inside the domain wall, is 10–100 times larger than in the domain. For a nucleus inside the domain wall, theory, in accordance with experiment, predicts a value of $\eta_{eff} \sim 10^3$–10^4 for the NMR amplification coefficient.

The power $\mathscr{P}$ absorbed by NMR is proportional to the square of the variable field amplitude, acting on the nuclear spins, i.e., in the case under consideration,

$$\mathscr{P} \sim \eta^2 h_x^2 . \tag{1.10}$$

Consequently, the amplification coefficient for the intensity level of the NMR signal will have the even larger value η^2 (or η_{eff}^2 for a nucleus inside domain walls).

Analogous amplification of NMR occurs in ferrites and also in antiferromagnets, in which the axis of antiferromagnetism lies in the "easy plane" (i.e., in the plane of smallest anisotropy). The problem of radiofrequency (r.f.) magnetic susceptibility and the NMR amplification coefficient will be elaborated in Chapter III.

A second group of effects is caused by the indirect interaction between nuclear spins via spin waves in a ferro- and an antiferromagnetic electron system.

The fact is that in magnetically ordered crystals side by side with the in-phase precession of all magnetic moments (uniform type of precession), there exist also weakly damped oscillation modes constituting a spatially nonuniform precession of the magnetic moments. Such nonuniform types of precession—waves of magnetization oscillation—are called *spin waves*. In quantum mechanics these waves are treated as some quasiparticles, the magnons, in analogy with the way the lattice vibrations are described in terms of quasiparticles, the phonons. The spin waves interact with nuclear spins as a result of the hyperfine coupling.

Let us consider a pair of nuclear spins I_1 and I_2. Each of these couples to the spin waves via the HFI (given by (1.2) for the magnetic atoms and by (1.2a), for instance, for nonmagnetic atoms), and thus I_1 and I_2 couple to each other in an indirect way. In the language of quantum mechanics this interaction between nuclear spins is caused by virtual processes of emission of spin waves (magnons) by one nucleus and their subsequent absorption by the other. This process leads to an interaction of the form (Suhl, 1958; Nakamura, 1958)

$$U(R_{12}) I_1^- I_2^+ \tag{1.11}$$

for the transverse components of the nuclear spins $I^\pm = I^x \pm I^y$, where $U(R_{12})$ is the parameter of this interaction depending in a definite way on the distance between the nuclear spins R_{12}. It is called the Suhl–Nakamura interaction.

Of course, there is an interaction between the longitudinal components also, but this is considerably less than the interaction (1.11) except at temperatures very close to the Curie (Néel) temperature. This means that the Suhl–Nakamura interaction is anisotropic in the sense that it depends on the direction of the nuclear spins relative to a physically distinguished direction Z (determined by the direction of the magnetization or the antiferromagnetic axis in ferro- and antiferromagnets respectively).

Note that this feature actually distinguishes the Suhl–Nakamura interaction from the analogous indirect interaction between nuclear spins via the conduction electrons (Ruderman and Kittel, 1954) which has the form of a scalar product

$$U_{RK}(R_{12})\, I_1 I_2 . \tag{1.12}$$

In contrast to (1.11) the interaction (1.12) depends only on the mutual orientation of the nuclear spins I_1 and I_2 and does not depend on their orientation relative to the axis of quantization.

The Suhl–Nakamura interaction leads to two directly observable effects. The first is obvious: the interaction given by (1.11) leads, because of its nonscalar character, to a nonzero second moment of the NMR line and thus contributes to its width. At a sufficiently high concentration of magnetic nuclei this broadening mechanism is apparently, in a number of cases, one of the basic broadening mechanisms.

The second effect is more subtle. It is associated with the correlation of the nuclear spin movements over distances of the order of the effective radius of interaction (1.11) and it consists of a shift in the NMR frequency at very low temperatures in comparison with its value ω_n at moderate temperatures. This dynamical shift is proportional to the mean value of the nuclear magnetization m at the given temperature, and its magnitude increases with decreasing temperatures, since $m \sim 1/T_n$ (where T_n is the temperature of the nuclear spin system) (De Gennes et al., 1963).

The inverse effect of a shift in the electron resonance frequency also exists. It is caused by the effective field

$$H_m = - \frac{\partial \mathscr{H}_{\mathrm{HFI}}}{\partial M} = - A_0 m , \tag{1.13}$$

applied to the electron spins by the nuclear spins (Heeger et al., 1961).

The above mentioned low temperature shift in the resonance frequency is particularly large for antiferromagnets with anisotropy of the "easy plane" type, such as for instance $MnCO_3$ and $CsMnF_3$, and in general, also for weakly anisotropic (cubic) antiferromagnets, such as $RbMnF_3$ and $KMnF_3$. In these antiferromagnets at temperatures of the order of $1°K$, the dynamical coupling between vibrations of the electron and the nuclear spins becomes so strong that it can in fact be considered as a single vibration of the electron-nuclear spin system as a whole (De Gennes et al.,

1963; Turov and Kuleyev, 1965). Since the parameter which determines the strength of the coupling is again the mean magnetization $m(T_n)$, it can be controlled by means of the NMR saturation effect, whereby the temperature of the nuclear spin system T_n can be varied over a wide range, from the temperature of the lattice up to ∞. Here a number of interesting nonlinear effects can be observed: the dependence of the frequency and the shape of both resonance lines on the intensity of the radio-frequency field, electron-nuclear double resonance, etc. (Heeger et al., 1961; Lee et al., 1963; Borovnik-Romanov and Tulin, 1965).

The nuclear spins in ferro- and antiferromagnets "see" not only the mean HFI field and its dynamical component, induced by the variable radiofrequency field, but also the thermal fluctuations of this field. The latter play a considerable role in the NMR relaxation processes (Reviews: 1962, 1965a, 1965b). At sufficiently low temperatures relative to the Curie (Néel) temperature many of these processes can be considered as the scattering of spin waves at nuclei (with or without the participation of phonons). The presence and the active role of the spin waves constitutes also the basic feature of NMR relaxation processes in ferro- and antiferromagnets. Chapter V of this book is devoted to this problem.

3. Basic Applications

Let us now turn to the information which can be obtained on ferro- and antiferromagnetism by means of NMR. As has already been mentioned at the beginning, the NMR experiments usually determine first of all the magnitude (and sometimes also the sign) of the local magnetic field at certain (magnetic) nuclei. From the distribution of the local field in the crystal one can then learn about the distribution of the spin density (the local magnetization) of the electron system. The knowledge of the latter has great significance for the construction and the development of the theory of the magnetically ordered state of matter, for the clarification of the actual mechanisms of the exchange forces responsible for this ordering.

In particular, in some cases NMR allows a more precise determination of the actual form of the magnetic structure. For instance, it was by this method that the presence of weak ferromagnetism in the antiferromagnet NiF_2, caused by the deflection of the magnetic moments of the sublattices from their strictly antiparallel configuration, was confirmed (Schulman, 1961a,b). For a ferrimagnet of complex structure, manganese chromite $MnCr_2O_4$, the angle of the cone along which the magnetic moments were arranged in this compound was determined (Houston and Heeger, 1964).

In compounds the measurement of the local fields at magnetic and especially at nonmagnetic ions supplies valuable information on the participation in the chemical bond of different electron groups (Schulman and Sugano, 1963).

In alloys it is possible to trace the evolution of the cooperative phenomena of ferromagnetism or antiferromagnetism through the study of the concentration dependence of the spin density distribution. In dilute alloys NMR has great perspectives in the investigation of localized or quasilocalized magnetic states associated with impurities. This is important for the theory of such alloys, which has been developed intensively in the last few years (see, e.g., Collection, 1963). In particular, NMR is one of the most effective methods for the investigation of the nature of the gigantic magnetic moments (of the order of 10 Bohr magnetons per atom), which arise as a result of the dissolution of some paramagnetic transition atoms in a non-magnetic matrix.

Since in the absence of an external field the NMR frequency is determined according to (1.7) and (1.6) by the spontaneous magnetization (or the magnetization of the corresponding sublattice) $M(T) \equiv \langle M_z \rangle$ at the given temperature, i.e.,

$$\omega_{n0} = \gamma_n A_0 M(T), \tag{1.14}$$

NMR is employed in a number of works as a method for the accurate measurement of the temperature dependence $M(T)$.

To be more precise, when there are several nonequivalent nuclei in the crystal, the temperature dependence of the mean magnetic moment of each species of atoms (ions) for the nuclei of which the NMR is observed can be determined from the corresponding resonance frequency ω_{n0}. The NMR frequency can be expressed in terms of "atomic" quantities in the form

$$\omega_{n0} = \frac{A\sigma}{\hbar}, \tag{1.14a}$$

where $\sigma = \langle S^z \rangle$ is the mean spin per atom of the given species. Accordingly, the frequencies ω_{n0} (and their temperature dependence) can differ not only for different nuclei but also for identical nuclei which are situated at lattice sites having different crystallochemical properties.

In particular, for impurity nuclei $\omega_{n0}(T)$ supplies "local" information on the temperature dependence of the magnetic moment localized near these nuclei. The latter can differ appreciably from the temperature dependence of the spontaneous polarization of the matrix (Jaccarino et al., 1966). These measurements allow the evaluation of the exchange interaction parameters for various pairs of atoms in alloys.

In the low temperature region the measurement of $M(T)$ or $\sigma(T)$ has for its object the verification of spin wave theory (Review, 1965a). This is not only a question of the temperature dependence $M(T)$ predicted by this theory, but also of what is essentially a quantum effect caused by the zero point quantum oscillations of the spins (in ferrites and antiferromagnets $\langle S^z \rangle - S \neq 0$ even for $T \to 0$).

At high temperatures the measurement of $M(T)$ in the direct vicinity of the Curie

(Néel) point T_C (T_N) is of great interest. Such measurements have great significance for the theory of phase transitions of the second kind.

Let us remark that in some antiferromagnets at very low temperatures the simple relation (1.14a) between ω_{n0} and σ can become more complex as a result of the above mentioned dynamic temperature shift, which is brought about by the Suhl–Nakamura interaction between nuclear spins.

And finally, still another, the last (in order but not in significance) application of NMR in ferro- and antiferromagnets which we shall mention: the study by NMR methods of the domain structure and the domain walls. The knowledge of the spin density distribution in the transition region between domains and of the spin motion in it, which can be obtained through NMR, plays a paramount role in the investigation of the physics of magnetization processes.

Intensive growth of work on NMR in ferromagnets started in 1960 following the discovery by Gossard and Portis (1959) of the unexpectedly large NMR signal in fine-grain powder specimens of cobalt. The authors themselves gave the correct interpretation of this effect of NMR amplification, based on ideas described above. In this work, as well as in the overwhelming majority of subsequent work, various resonant signals were observed from nuclei lying in the domain walls of ferromagnets. They have a considerably higher intensity than signals from nuclei in the domains and in single-domain specimens, and require much simpler apparatus for their investigation. The interpretation of the results stimulated further development of the theory of domain walls, taking into consideration their special structure and requiring a more detailed and deeper physical approach (Winter, 1961).

In the following the basic theoretical ideas on NMR phenomena in ferromagnets and antiferromagnets are set forth. The experimental material is on the whole presented only by way of illustration. For a more detailed study of experimental results, the reader is referred to the review articles of Narath (Review, 1967), Jaccarino (Review, 1965a), Portis and Lindquist (Review, 1956) and also of Winter (Review, 1962). In Appendix II we present as reference a table of experimental data the basis of which was borrowed by us from the reviews mentioned and which was supplemented by new data taken from original articles.

CHAPTER II

Resonance Frequencies

1. Resonance Frequencies and the Spin Density Distribution

As has already been mentioned (Chapter I), NMR is one of the effective methods for the study of spin density distribution in magnetically ordered crystals. More precisely, NMR supplies information on the distribution of magnetic moment in a crystal, since the local fields at the nuclei may be both of spin and orbital origin.

We shall right away point out two aspects of such investigations. First, in many cases NMR enables one to find the mutual orientation of the magnetic moments of various groups of atoms (magnetic sublattices) in relation to each other, as well as their orientation with respect to the crystallographic axes. This method can be of considerable help in determining (or refining) the type of magnetic structure. The spectrum of resonance frequencies will appreciably depend on the character of the magnetic order in the crystal of interest: whether this is a simple ferromagnet (i.e., a ferromagnet with a single magnetic sublattice) or a collinear ferrimagnet (i.e., a ferromagnet with two or more magnetic sublattices, the moments of which are oriented antiparallel to each other), whether it is an antiferromagnet with fully compensated magnetic moment or an antiferromagnet with weak magnetism; or, finally, whether this is a magnet with a complicated noncollinear (e.g., spiral) magnetic structure. A more detailed treatment of the problem of NMR frequency spectra for the principal types of magnetic structures constitutes the main contents of the present chapter. In general, the discussion will center on ferromagnets, antiferromagnets and ferrimagnets with collinear (or weakly noncollinear) magnetic structure, for which all the magnetic moments are directed along or against the axis of magnetic ordering.

The second aspect of the investigation of NMR frequencies is connected with

the study of the local spin density distribution, its sign and magnitude and, finally, its origin.

In this connection the following problems are investigated: NMR frequencies in dilute alloys, the hyperfine shift of NMR frequencies in chemical compounds at the nuclei of nonmagnetic ions, the influence of various crystal defects on the resonance. In this chapter we shall mainly consider the resonance of the nuclei of magnetic atoms or ions, while nonmagnetic ions will be dealt with in Chapter VIII, which is dedicated to this subject. The inhomogeneity in the distribution of magnetization, caused by the existence of domain structure in ferro- and antiferromagnets, will also be excluded from consideration for the time being. The special features of NMR for nuclei in domain walls will be discussed in Chapter VII.

If correlation effects caused by indirect interaction between nuclear spins via the electron spin system (see Section 3) are neglected, the NMR frequency will be determined by the mean value, at the given temperature, of the resultant local magnetic field H_{loc} at the nucleus of interest[*]

$$\omega_n = \gamma_n \left| \langle H_{\text{loc}} \rangle \right|. \tag{2.1}$$

Expression (2.1) does not take into consideration quadrupole splitting, which appears for nuclear spins $I > 1/2$ whose surroundings do not possess cubic symmetry. Quadrupole effects will be considered in Chapter VI.

The local field can be represented as a sum of the hyperfine and the dipole terms plus the external field:

$$\langle H_{\text{loc}} \rangle = H_n + H_{\text{dip}} + H. \tag{2.2}$$

Here

$$H_n = -A_0 \langle M \rangle \equiv -\frac{A\sigma}{\gamma_n \hbar}, \tag{2.3}$$

$$H_{\text{dip}} = \sum_j \left(\frac{\langle \mu_j \rangle}{R_j^3} - 3 \frac{R_j (\langle \mu_j \rangle R_j)}{R_j^5} \right), \tag{2.4}$$

$\langle \mu_j \rangle$ is the magnetic moment of the atom (ion) situated at a point given by the radius vector R_j (the nucleus is taken as the origin of the coordinate system).

Let us again mention that $\langle M \rangle$ is the local magnetization, equal to the magnetic moment of the atom which contains the nucleus of interest divided by the appropriate atomic volume. For nuclei of impurity atoms and also, in general, for nuclei of matrix

---

[*] The fluctuations of H_{loc} may, in general, produce a shift in the resonance frequency proportional to the square of the HFI constant; in the majority of cases, however, this shift too can be neglected (Turov et al., 1967; Kurkin and Turov, 1967). This problem will be discussed in more detail in Chapter V, Sec. 5.

atoms close to them, it is necessary to take into account the difference between $\langle M \rangle$ and the saturation magnetization (in case of a simple ferromagnet) or the magnetization of the corresponding sublattice (in case of a ferrimagnet or an anti-ferromagnet). This factor is emphasized in the right-hand side of equation (2.3) which expresses H_n in terms of the mean spin $\sigma = \langle S \rangle$ (or the angular momentum $\langle J \rangle$) localized in the vicinity of the given nucleus.

The contribution to H_{dip} from magnetic moments μ_j removed far enough from the nucleus can be taken into account by a continuous medium model, replacing the summation in (2.4) by integration. This part of H_{dip} is made up of the Lorentz field $(4\pi/3) M_{\mathrm{spec}}$ and the depolarization field of the body surface $-\overset{\leftrightarrow}{N} M_{\mathrm{spec}}$, where $\overset{\leftrightarrow}{N}$ denotes symbolically the tensor of the depolarizing factors $N_{\alpha\beta}$, and M_{spec} is, in general, the total magnetic moment of the specimen, related to unit volume. Taking into consideration the discrete character of the medium leads, in the case when the symmetry of the atomic arrangement in the vicinity of the nucleus (inside the so-called Lorentz sphere) is lower than cubic, to an additional contribution (H_{sph}) to the total dipole field, depending on the actual arrangement of the atoms and the direction of their magnetic moments. Thus

$$H_{\mathrm{dip}} = \frac{4\pi}{3} M_{\mathrm{spec}} - \overset{\leftrightarrow}{N} M_{\mathrm{spec}} + H_{\mathrm{sph}} \tag{2.5}$$

(Robert and Hartmann-Boutron, 1962; White, 1965).

In particular, for spherical specimens, for which $N_{\alpha\beta} = (4\pi/3) \delta_{\alpha\beta}$ (where $\delta_{\alpha\beta} = 1$ when $\alpha = \beta$ and $\delta_{\alpha\beta} = 0$ when $\alpha \neq \beta$), the Lorentz field and the demagnetization field fully cancel each other.

It must be borne in mind that for nuclei where the local symmetry of the surroundings is lower than cubic, the HFI parameter A too is, in general, a tensor of the corresponding symmetry. For nuclei of magnetic atoms, however, this anisotropy of the HFI is, as a rule, relatively small.

Since both H_{dip} and H are much smaller than H_n, in the first approximation

$$\omega_n \simeq \gamma_n \left| H_n + H_{\mathrm{dip}}^z + H^z \right| \tag{2.6}$$

where the direction of the vector M was taken as the Z axis. As has already been mentioned in Chapter I, in the majority of cases

$$H_n = - A_0 M$$

is negative. (Henceforth we omit the "mean" sign in $\langle M \rangle$.)

For nuclei having a cubic surrounding in ferro- and antiferromagnets magnetized to saturation $(M \parallel H)$, we have instead

$$\omega_n \simeq \gamma_n \left| H_n + \left(\frac{4\pi}{3} - N_{zz} \right) M_s + H \right|, \tag{2.7}$$

where M_s is the saturation magnetization. If we are not dealing with impurity nuclei (and, in general, not with their nearest neighbors) then for simple ferromagnets M_s is identical with M, for collinear ferrimagnets M_s is equal to the algebraic sum of the magnetizations of the sublattices. For instance, for a ferromagnet with two sublattices having oppositely directed magnetic moments, $M_s = M_1 - M_2$. In the latter case we have two NMR frequencies

$$\omega_{n1} = \gamma_n \left| A_{10}M_1 - H - \left(\frac{4\pi}{3} - N_{zz} \right)(M_1 - M_2) \right|,$$

$$\omega_{n2} = \gamma_n \left| A_{20}M_2 + H + \left(\frac{4\pi}{3} - N_{zz} \right)(M_1 - M_2) \right|,$$

(2.8)

where A_{10} and A_{20} are HFI constants for the nuclei of the first and the second sublattice respectively.

In a particular case ($M_1 = M_2 = M_0$ and $A_{10} = A_{20} = A_0$) expressions (2.8) are valid also for antiferromagnets, if the magnetic field is directed along the axis of antiferromagnetism (i.e., the axis parallel and antiparallel to which the magnetic moments in the antiferromagnet are ordered*). When H is not parallel to the anti-ferromagnetic axis and is not very large (again $H \ll |H_n|$ and, in addition, $H \ll H_E$, where H_E is the effective field of the exchange forces responsible for antiferromagnet-ism; see Section 3), H is replaced by its projection on this axis.

For nuclei with noncubic surroundings the calculation of the dipole field com-ponent H_{sph}^z requires knowledge of the actual crystal and magnetic structures, the magnitude of the magnetic moments and the interatomic distances for the nearest neighbor atoms. Let us remark that the local symmetry (the symmetry of the sur-roundings of the nucleus under consideration) can readily be noncubic even if the crystal is cubic. Thus, for instance, for the iron nucleus in yttrium ferrite garnet ($Y_3Fe_5O_{12}$) the crystal lattice of which belongs to the cubic system, the symmetry of the surroundings in the tetrahedral and octahedral positions can be described by the point groups $\bar{4}$ or $\bar{3}$, respectively (four-fold or three-fold inversion axes). The magnitude of H_{sph}^z and its dependence on the direction of M for these positions are evaluated in the work of Boutron and Robert (1961). It is important to note that H_{sph}^z can differ even for crystallographic positions of the same symmetry. There are three types of tetrahedral positions corresponding to the three possible orienta-tions of the $\bar{4}$ axis (along the three edges of a cube) and four types of octahedral positions corresponding to the four possible orientations of the $\bar{3}$ axis (along the four diagonals of a cube). For this reason the field H_{sph}^z can produce additional splitting of the NMR line.

---

* To be more precise, when the nonzero parallel susceptibility of the antiferromagnet $\chi_{\parallel}$ at $T \neq 0$ is allowed for, then $M_1 = M_0 + \chi_{\parallel}H/2$ and $M_2 = M_0 - \chi_{\parallel}H/2$.

If in the near vicinity of the nucleus of interest there is an impurity atom with a magnetic moment which differs by $\Delta\mu$ from the moment of the matrix atoms, then naturally this will also destroy the cubic symmetry and lead to the appearance of an additional dipole field at the nucleus (Murray and Marshall, 1965):

$$H^z_{sph} = \frac{\Delta\mu}{R^3}(3\cos^2\theta - 1),\qquad(2.9)$$

where R is the radius vector from the nucleus to the impurity and θ is the angle which it makes with the Z axis. Such a field must be added to expression (2.7) for a nucleus near an impurity.*

From the measurement of NMR frequencies as a function of the external field H and the use of expression (2.6) or (2.7) it is possible to determine not only the magnitude but also the sign of the local field at the nucleus (by extrapolation to $H = 0$), while from the slope factor of the line $\omega_n = \omega_n(H)$ we can also find the gyromagnetic ratio γ_n and, consequently, the magnetic moment of the nucleus $\gamma_n \hbar I$, if its spin I is known. An example of such a measurement is shown in Figure 3 which presents

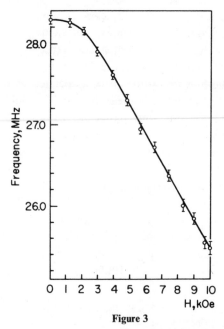

Figure 3

The dependence of the NMR frequency of the Ni[61] nucleus in metallic nickel at 77°K on the intensity of the external magnetic field H (Streever, 1963).

---

* Here we shall disregard for the time being other effects, caused by the impurities, which lead to a redistribution of the electron and spin densities at neighboring nuclei and thus cause a change in the local field.

the dependence of ω_n on H for the Ni^{61} nucleus in metallic nickel at 77°K (Streever, 1963). From the linear part of the curve in Figure 3 the true magnitude of the magnetic moment of Ni^{61}, $\mu(Ni^{61}) \simeq 0.70$ nuclear magnetons, was first determined, as well as the hyperfine field, $H_n \simeq -80$ kOe (at 77°K).

Recently, very many NMR investigations are being carried out in dilute alloys.* The detailed study of the spatial distribution of spin polarization around the impurity atom, which is obtained from the hyperfine structure of the NMR lines, has considerable significance for the solution of such fundamental problems as the nature of localized magnetic moments in metals and alloys and the nature of the ordering of these moments in the ferromagnetic (or antiferromagnetic) state.

The NMR fine structure in dilute alloys (at matrix nuclei) most often appears in the form of satellites at the wings of the fundamental line. The reasons for the appearance of satellites can be the following.**

One reason is the difference in the magnetic moments of the impurity atoms and the matrix, which leads to an additional dipole field at the matrix nuclei determinable by expression (2.9). The magnitude of the field H^z_{sph} will differ for various radius vectors R (not only for different coordination spheres whose radius is determined by the magnitude of R, but, in general, also for various angles θ in the same sphere). In the vicinity of an impurity, the spectrum of the fields H^z_{sph} will essentially be discrete, which will lead to the appearance of satellites (provided, of course, the shift in the NMR line caused by these fields is not less than the line width).

The second cause of the fine structure can be associated with the redistribution of the electron and spin densities localized near the matrix nuclei in the vicinity of the impurity atom. The latter leads, first, to a change in HFI, which depends in the main only on the distance R between the nucleus and the impurity and, second, to a change in the dipole, quadrupole, etc., fields, which depend on the angle between R and the magnetization M.†

The investigation by NMR methods of dilute alloys of magnetic atoms (Fe and Co) in a nonmagnetic matrix (Pt or Pd), in which the so-called gigantic magnetic moments are detected, is of particularly great interest. These alloys are ferromagnetic at low temperatures and their saturation magnetization per magnetic atom reaches $10\mu_B$. The spatial distribution of the gigantic magnetic moment can be studied by NMR (Ehara, 1964; Kobayashi and Itoh, 1965).

---

 * See, e.g., the articles of Kushida et al. (1962), Portis and Kanamori (1962), Murray and Marshall (1965), Itoh et al. (1965), Jaccarino (1965, 1968).

** Here we are not taking into account for the time being the quadrupole splitting.

 † It must be admitted that for metals and alloys (in particular for the iron group) the division of the local field into hyperfine and dipole parts is in fact difficult, since not only the electrons of the outer shell (the "4s-electrons"), which are basically responsible for electrical conduction, but also the electrons of the inner, unfilled shell (the "3d-electrons"), which are responsible for ferromagnetism, show collective behavior to a degree. See, Review (1967).

For instance, in the alloy Pt + 1% Fe the resonance of the Pt nucleus extends approximately from 100 to 600 MHz and has a number of maxima (Kobayashi and Itoh, 1965). The deciphering of the NMR spectrum shows that the region of spin polarization of the Pt atoms in the vicinity of the impurity Fe atom extends over at most a few lattice distances, as the magnetic moments of the Pt atoms decrease fast with their distance from the impurity. Each maximum corresponds to a group of Pt atoms situated at equal distances from the Fe, i.e., nearest neighbors, next-nearest neighbors, etc.

The satellites can also be observed for the resonance of the impurity nucleus itself (La Force et al., 1962). In particular, in the alloy Fe + 1% Co at 24°C there are satellites at much higher frequencies, 282.3 and 285.6 MHz, than the frequency of the main Co^{59} maximum at 278.8 MHz. The main maximum is ascribed to a Co nucleus having as its nearest neighbors eight Fe atoms, the second maximum to a Co nucleus with seven Fe and one Co neighbors and, finally, the third maximum is associated with a Co nucleus with six Fe and two Co neighbors, etc. A shift of 3.4 ± 0.1 MHz in the resonance frequency is obtained by each substitution of a cobalt atom for a nearest iron atom.

The alloy V–Fe (from 0 to 34% iron) serves as a good example which shows that it is possible by the use of NMR to trace the appearance of localized moments in an alloy and, subsequently, with the increase in concentration, to detect cooperative phenomena (ferromagnetism). The measurement of the static magnetic susceptibility and resonance of the V^{51} nucleus indicates the appearance of the d-electron spin polarization near the iron atoms only for a composition of about 23% Fe (Lam et al., 1963). Below this concentration the d-electrons are in an energy band with paired spins. Ferromagnetism does not appear as long as the iron content does not reach 27%.

A summary of experimental data for resonance frequencies and local fields is given in Appendix II.

2. Temperature Dependence of the NMR Frequency

The investigation of the temperature dependence of magnetic moments in magnetic crystals constitutes the widest and most effective application of NMR. This application is associated with the fact that the local magnetic field at the nuclei and, consequently, the resonance frequencies are determined by the mean value of the local magnetization at the given temperature.* Therefore, the measurement of the temperature dependence of NMR frequencies gives also the temperature dependence

---

* The hyperfine field at the nuclei of nonmagnetic ions and also the dipole field are determined by the mean magnetic moments of the surrounding magnetic ions.

of the local magnetization. This information is necessary for the verification and further development of spin wave theory, and also for the construction of a theory of ferro- and antiferromagnetic phase transitions.

The advantage of the NMR method in the investigation of $M(T)$ consists, first of all, in its high accuracy. In the case of sufficiently narrow NMR lines the relative accuracy of the magnetization measurement can reach 10^{-5}. Besides, this method allows the investigation of the temperature dependence of magnetization of separate sublattices (in ferrimagnets and antiferromagnets) and even of individual atoms (for example, for impurities).

Let, for instance, the dipole contribution to the NMR frequency be insignificant. Then for $H \to 0$

$$\omega_n(T) = \omega_{n0}(T) = \gamma_n A_0 M(T), \tag{2.10}$$

and if the HFI constant A_0 itself does not depend on temperature, the temperature variation of $M(T)$ will directly determine the variation of $\omega_n(T)$. Actually, there are experimental indications to the effect that in some cases A_0 too possibly depends on the temperature (Benedek and Amstrong, 1961). Apparently, in the majority of cases the temperature dependence of A_0 can, however, be neglected.

In this connection it must be borne in mind that, strictly speaking, the relation (2.10) describes the dependence of ω_n on T at constant volume (Robert and Winter, 1960). Consequently, its use assumes the exclusion of thermal expansion. The latter can be done by a thermodynamic recomputation if, in addition, also $(\partial \omega_n / \partial V)$ is measured and the coefficient of thermal expansion $(1/V)(\partial V/\partial T)_P$ is known (Benedek and Amstrong, 1961; Heller, 1966).

For simple ferromagnets with a single sublattice, and also, in general, for more complex structures having completely equivalent magnetic sublattices with the same temperature dependence of their magnetizations (for example, for antiferromagnets), the introduction of the dipole field does not change the direct proportionality between $\omega_n(T)$ and $M(T)$. At the same time, for ferrimagnets with nonequivalent magnetic sublattices, the resonance frequency of one sublattice will be influenced not only by the magnetization of the same sublattice, but, as a result of the dipole interaction, partially also by the magnetization of the other sublattices. All this is also true for the HFI if the nuclei belong to nonmagnetic atoms.

Low Temperatures and the Theory of Spin Waves

The aim of many works on the investigation by the NMR method of the temperature dependence of the local magnetization $M(T)$ is the confirmation of conclusions which follow from the theory of spin waves in the region of low temperatures (relative to the Curie temperature T_C in ferromagnets, or the Néel temperature T_N in antiferromagnets). We shall therefore first dwell briefly on the theory of spin waves. In magnetically ordered crystals, alongside the waves of lattice vibrations, the

phonons, there may also be excited propagating waves of magnetic moment vibrations, the spin waves or magnons, first considered by Bloch (1930).

The thermal excitation of spin waves produces some destruction of the magnetic order as a result of the random deflections of magnetic moments from their equilibrium directions, just as the thermal vibrations of the lattice displace its atoms from their equilibrium positions in a random manner. There are, however, essential differences between lattice vibrations and magnetic moment vibrations. While the lattice vibrations do not change in the first (harmonic) approximation the time average of the atomic location, the magnetic moment vibrations in the same approximation do change (namely decrease) the observed mean magnitude of the magnetic moment (i.e., the time average of the projection of each moment on its equilibrium direction). This effect is revealed by a shift in the NMR frequency.

Let us consider a ferro-, antiferro- or ferrimagnet, in which the magnetic moments of the sublattices are directed along the appropriate "easy axes." Let the directions of the "easy axes" of the various magnetic sublattices differ only as to their sign, i.e., all axes are directed along or against the same z axis of the crystal (collinear magnetic structure) and let an external field be also applied parallel to the same axis.

The spin Hamiltonian of such a system can be written in the form

$$\mathcal{H} = -\sum_{j,j'} J(R_{jj'}) S_j S_{j'} - \hbar \sum_j \gamma_j S_j^z H_j + \mathcal{H}'_{\text{dip}}. \qquad (2.11)$$

Here the first sum represents the exchange interaction, the parameters of which $J(R_{jj'})$ (the exchange integrals) depend on the distance $R_{jj'} = |R_j - R_{j'}|$ between the spins S_j and $S_{j'}$*. The second sum describes that part of the magnetic energy which can be calculated by the introduction of the effective internal magnetic field $H_j \| Z$, acting on the spin S_j. The field H_j is composed of the external field H, the demagnetization field of the specimen surface $-N_{zz} M_{\text{spec}}$ and the effective field of magnetic anisotropy H_a^j, which has the same magnitude for sites j belonging to the same magnetic sublattice but differs in sign for crystallographically equivalent magnetic sublattices with oppositely directed magnetizations; $\gamma_j = g_j e/2mc$ is the gyromagnetic ratio for the spins of the respective magnetic sublattice ($\gamma_j < 0$, since the charge of the electron $e < 0$). Finally, the last term $\mathcal{H}'_{\text{dip}}$ represents the usual dipole–dipole interaction of the magnetic moments, with the exclusion of that part which has already been taken into account via H_j (in the form of the energy of the surface demagnetization field, and also, in general, the energy of magnetic anisotropy).

Let us introduce the quantity ζ_p determining the sign of the spin projection for the p-th magnetic sublattice: $\zeta_p = +1$ for a sublattice with spins down (and, consequently, with magnetization directed "upward"), $\zeta_p = -1$ for a sublattice with

---

* The exchange interaction, described in the form of a sum of scalar products of spin operators with coefficients depending only on the distance between the respective spins, is called the Heisenberg exchange interaction.

spins up. According to Holstein and Primakoff (1940) it is then possible to introduce for each of the sublattices operators $a_p^+(R_j)$ and $a_p(R_j)$ (the so-called spin deflection creation and annihilation operators) through which the spin operators S_j can be approximately expressed in the following form:

$$\left.\begin{array}{l} S_{j(p)}^+ \simeq (2S_p)^{1/2} a_p^+(R_j), \\ S_{j(p)}^- \simeq (2S_p)^{1/2} a_n(R_j), \end{array}\right\}, \quad \text{when } \zeta_p = -1,$$

$$\left.\begin{array}{l} S_{j(p)}^+ \simeq (2S_p)^{1/2} a_p(R_j), \\ S_{j(p)}^- \simeq (2S_p)^{1/2} a_p^+(R_j), \end{array}\right\}, \quad \text{when } \zeta_p = +1.$$

(2.12)

Finally,

$$S_{j(p)}^z = \zeta_p \left[S_p - a_p^+(R_j)\, a_p(R_j) \right]. \tag{2.12a}$$

Here, $S^\pm = S^x \pm iS^y$, the index p indicates that the cell number index j refers only to the p-th magnetic sublattice, consisting of the spins S_p.

We can express the Hamiltonian (2.11) in terms of the operators a_p^+ and a_p with the use of (2.12), and subsequently discard in it all terms of higher than second order in these operators. This means that we are considering only the harmonic vibrations of the spins (in analogy with the treatment of harmonic lattice vibrations where terms higher than second order in strain were dropped from the expression for the elastic energy). The quadratic form in the operators $a_p^+(R_j)$ and $a_p(R_j)$ thus obtained will contain, in particular, products of operators related to different lattice cells j (and also, in general, to different p), i.e., nondiagonal terms of the form $a_{p'}^+(R_{j'})a_p(R_j)$, etc. This means that a spin deflection $S_j - S_j^z$ arising at some lattice cell j can propagate over the whole crystal (as the operator $a^+(R_{j'})\, a(R_j)$ annihilates such a deflection at the cell j and recreates it at the cell j'). Any spin deflection belongs, therefore, in fact to the whole crystal, and for its description p and R_j should be replaced with some other, collective variables (such as, for example, a wave vector). This can be achieved by a so-called unitary transformation from the operators $a_p^+(R_j)$ and $a_p(R_j)$ to new operators $a_v^+(k)$ and $a_v(k)$, indexed by the new (collective) quantum numbers v and k, such that the quadratic form will be diagonal (only terms of the form $a_v^+(k)\, a_v(k)$ will be present in it). Following this transformation the Hamiltonian (2.11) becomes the sum of the ground state energy and the energy of elementary excitations, the spin waves (magnons):

$$\mathcal{H} = \mathcal{H}_0 + \Delta\mathcal{H}_0 + \sum_{vk} \varepsilon_v(k)\, n_{vk}. \tag{2.13}$$

Here $n_{vk} = a_v^+(k)a_v(k)$ is the number operator for spin waves of "species" v with wave vector k, $\varepsilon_v(k)$ is the energy of one spin wave in this state (vk). The number of spin wave branches ("species") equals, in the general case, the number of magnetic sublattices, describing the magnetic structure of the crystal, while the number of

values of the quasicontinuous wave vector k equals the number of magnetic unit cells in the sample.

The operators $a_v^+(k)$ and $a_v(k)$ describe the creation and annihilation respectively of spin waves in the (vk) state. They satisfy the following commutation relations:

$$a_v(k)a_{v'}^+(k') - a_{v'}^+(k')a_v(k) = \delta_{vv'}\delta_{kk'}$$
$$a_v(k)a_{v'}(k') - a_{v'}(k')a_v(k) = 0, \text{ etc.,} \tag{2.14}$$

and consequently, the spin waves are quasiparticles of the Bose type.

It was just the noncommutativity of the operators $a_v^+(k)$ and $a_v(k)$ that led to the appearance of the term $\Delta\mathcal{H}_0$ in (2.13). This is the energy of the so-called zero-point vibrations. The term $\mathcal{H}_0$ is the energy of the nominal ground state, i.e., the energy of that strictly ordered arrangement of spins, a priori assumed by us, according to which in each site of sublattice p the spin has a strictly determined z projection, equal to its nominal value $\zeta_p S_p$. The renormalization of the ground state energy brought about by the term $\Delta\mathcal{H}_0$ means in fact that not only the spin of a single site but also, in general, the total spin of each of the sublattices may not have a strictly determined value.* We can only speak of the mean (expectation) value of the spin $\sigma_p = \langle S_p^z \rangle$, which even at the absolute zero temperature (i.e., even in the total absence of spin waves, when all $n_{vk} = 0$) may prove not equal to S_p in magnitude (Anderson, 1952; Kubo, 1952).

The situation mentioned is characteristic in the main of ferrimagnets and, in particular, of antiferromagnets which have sublattices with oppositely directed spins (precisely in such magnets the spin of a separate sublattice is not a conserved quantity because of its exchange interaction with other sublattices).

Today the NMR method is the only method capable of measuring with sufficient accuracy the deviation of the mean spin $|\sigma_p|$ at $T = 0$ K from its nominal value S_p, which is a significant factor for the problem of the ferri- and antiferromagnetic ground state. Since at $T = 0°$K the spontaneous magnetization of sublattice p can be computed from the expression

$$M_p(0) = -\frac{\partial}{\partial H_p}(\mathcal{H}_0 + \Delta\mathcal{H}_0)\frac{\partial H_p}{\partial H}\bigg|_{H=0} \qquad (H_p \equiv H_{j(p)}), \tag{2.15}$$

the change of spin predicted by the theory of spin waves, due to the zero point vibrations, is determined exactly by the correction $\Delta\mathcal{H}_0$ to the ground state energy (Bogolyubov and Tyablikov, 1949; Tyablikov, 1965),

$$\Delta S_p(0) = \zeta_p S_p - \sigma_p = -(\gamma_p N_p \hbar)^{-1}\frac{\partial\Delta\mathcal{H}_0}{\partial H_p}\bigg|_{H=0} \tag{2.16}$$

---

* This follows from the fact that in the general case the operator of such a total spin $\sum_{j(p)} S_j^z$ does not commute with the Hamiltonian (2.11).

(where N_p is the number of spins per unit volume in the sublattice p; we have taken into account that $\partial H_p/\partial H = 1$). The ratio $\Delta S(0)/S$ is of the order of $(1/zS)$, where z is the number of nearest neighbors (Pu Fu-cho, 1960; Oguchi, 1960).

It is interesting that existing experimental data for antiferromagnetic manganese compounds such as MnF_2 and $KMnF_3$ allegedly indicate that the theory of spin waves leads to an increase in the value of $\Delta S/S$ (Jones and Jeffers, 1964).*

The temperature dependence of the magnetization $M_p(T)$ is determined by the mean number of spin waves $\bar{n}_{vk} \equiv \langle n_{vk} \rangle$ at the given temperature which for Bose quasiparticles is given by the Bose–Einstein distribution function

$$\bar{n}_{vk} = \exp[\varepsilon_v(k)/\kappa T - 1]^{-1}. \tag{2.17}$$

The thermodynamic potential of such a mixture of ideal gases consisting of Bose quasiparticles (magnons) with energy $\sum_{vk} \varepsilon_v(k)\bar{n}_{vk}$ equals as known (Landau and Lifshitz, 1964)

$$\Omega_M = \kappa T \sum_{vk} \ln[(1 - \exp(-\Sigma_v(k)/\kappa T)]. \tag{2.18}$$

Therefore the change in the magnetization of the p-th sublattice, caused by spin waves, can be written in the form

$$\Delta M_p(T) \equiv M_p(0) - M_p(T) = \frac{\partial \Omega_M}{\partial H_p} \frac{\partial H_p}{\partial H} = \sum_{vk} \left(\frac{\partial \varepsilon_v(k)}{\partial H_p} \bar{n}_{vk} \right). \tag{2.19}$$

The number of spin wave branches (i.e., the number of values taken by the index v) and the explicit dependence of $\varepsilon_v(k)$ on the wave vector k, the parameter H_p, etc., is determined for each branch by the actual type of magnetic structure. The total number of branches equals the number of magnetic sublattices n; not every branch, however, contributes to $\Delta M_p(T)$ at low temperatures. Only the spin waves with a sufficiently small energy gap ε_{v0}, the so-called acoustic type spin waves,** contribute to $\Delta M_p(T)$.

Ferro- and Ferrimagnets

In the case of ferro- and ferrimagnets there is one acoustic branch (corresponding to the in-phase vibration of the magnetic moments of all the sublattices) and $n - 1$ branches of optical type (corresponding to the vibration of the magnetic moments of some sublattices in counterphase with that of others). The energy gap for the acoustic branch is determined only by the magnetic interactions, since the motion involves

---

* For an explanation see, for example, Review (1967).

**The energy gap $\varepsilon_{v0} \equiv \varepsilon_v(0)$ is defined as the smallest energy necessary for the excitation of spin waves of a given species v. It is clear that spin waves for which $\kappa T \ll \varepsilon_{v0}$ at the temperature of interest cannot contribute to $\Delta M_p(T)$.

the vibration of the total magnetic moment of all the sublattices relative to the effective field H_p, while the angle between neighboring magnetic moments does not change. On the contrary, during optical vibrations the angle between the magnetic moments of different sublattices undergoes a change and this means that the energy gaps for the corresponding spin wave branches will be determined by the exchange interactions between the relevant magnetic sublattices. It is precisely for this reason that the optical branches are not excited at low temperatures and they can be neglected in the computation of $M_p(0) - M_p(T)$, in the first approximation.

As a result of the translational symmetry, associated with the periodicity of the magnetic structure, the energy of the acoustic spin waves $\varepsilon_a(k)$ (and, of course, also of the optical ones) will be a periodic function in the wave vector or k-space: $\varepsilon_a(k + K) = \varepsilon_a(k)$, where K is any vector of the magnetic reciprocal lattice. At low temperatures the thermal motion excites mainly vibrations with small k (such that $ak \ll 1$, where a is the interatomic distance) and, consequently, $\varepsilon_a(k)$ can be expanded in powers of ak. As a result, the energy of spin waves of the acoustic type in ferro- and ferrimagnets can be expressed approximately in the following manner

$$\varepsilon_a(k) = \varepsilon_k (1 + \Phi_k \sin^2\theta_k)^{1/2} . \tag{2.20}$$

Here

$$\varepsilon_k = J_{eff}(ak)^2 + \hbar|\gamma_{eff}|H_{eff} \tag{2.21}$$

is the spin wave energy when the inhomogeneous part of the dipole interaction is neglected (i.e., the term $\mathscr{H}_{dip}$ in expression (2.11)); the latter is taken into account by the second term in expression (2.20) with the multiplier

$$\Phi_k = \hbar|\gamma_{eff}| \cdot \frac{4\pi M_s}{\varepsilon_k} ; \tag{2.22}$$

θ_k is the polar angle of the vector k, measured from the z axis. J_{eff} is some effective parameter of the exchange interaction; for example, in a ferromagnet with a simple cubic lattice, when in (2.11) only the exchange interaction between nearest neighbors is taken into account, $J_{eff} = 2SJ$, where J is the "exchange integral" for these neighbors. Further, γ_{eff} is the mean gyromagnetic ratio, defined as the ratio of the total magnetic moment of all the sublattices ($M_s = \sum_p M_{p0}$) and their total angular momentum:

$$\gamma_{eff} = \frac{\sum_p M_{p0}}{\sum_p \hbar S_{p0}}, \tag{2.23}$$

where $M_{p0} = \hbar\gamma_p S_p$ and $S_{p0} = \zeta_p N_p S_p$. Finally, the effective field H_{eff} is equal to the average of the effective fields of all the sublattices

$$H_{\text{eff}} = \frac{\sum\limits_p M_{p0} H_p}{\sum\limits_p M_{p0}}. \tag{2.24}$$

If we ignore the contribution of the optical branches, expression (2.19), taking (2.20)–(2.24) into account, can be rewritten in the form

$$\Delta M_p(T) = \hbar |\gamma_{\text{eff}}| \frac{M_{p0}}{M_s} \sum_k \frac{1 + \frac{1}{2} \Phi_k \sin^2\theta_k}{(1 + \Phi_k \sin^2\theta_k)^{1/2}} \, \bar{n}_{ak}. \tag{2.25}$$

Thus, in the present approximation (neglect of the interaction between spin waves and the contribution of the optical branches; limitation of the exchange energy expansion in powers of k to terms not higher than quadratic) the relative changes in the magnetizations of all the sublattices are identical: $\Delta M_p(T)/M_{p0} = f(T)$ is a function of temperature, not depending on the index p of the sublattice.

Let us remark that expression (2.20) for the energy of spin waves, when applied to ferromagnets, and, consequently also expression (2.25) for $\Delta M(T)$, remain valid even for a more general form of the energy ε_k, represented by expression (2.21), without expanding its exchange part in powers of k. This more general form is the following:

$$\varepsilon_k = 2S \sum_{j(\neq j')} J(R_{jj'}) \left[1 - \exp(-jkR_{jj'}) \right] + \hbar |\gamma_e| H_{\text{eff}}. \tag{2.21a}$$

At the same time for ferrimagnets we have dropped in expression (2.21) (compared to unity) small terms of the order of the ratio $\hbar \gamma_p H_p / J_{\text{eff}}$, which, when taken into account, may in principle lead to a small difference in $\Delta M_p(T)/M_{p0}$ for different sublattices even in the approximations mentioned above.

Let us consider first the region of not very low temperatures, such that

$$\kappa T \gg \hbar |\gamma_{\text{eff}}| H_{\text{eff}} \quad \text{and} \quad \hbar |\gamma_{\text{eff}}| 4\pi M_s \tag{2.26}$$

(but, of course, still small in comparison with the Curie temperature). In this case, expression (2.25) leads to the well known "three halves law" for $\Delta M_p(T)$, first obtained by Bloch (1930) for ferromagnets:

$$\frac{\Delta M_p(T)}{M_{p0}} = C_{3/2} T^{3/2}, \tag{2.27}$$

where $C_{3/2}$ is a constant, not depending on the temperature.

The "three halves law" was verified experimentally as the first term in the expansion of $\Delta M(T)$ in powers of the temperature both by direct magnetic measurements and by the NMR method in a number of ferro- and ferrimagnets, in which it is applicable over a sufficiently wide temperature interval. As an example, Figure 4 shows the temperature dependence of the NMR frequency of the Mn^{55} nucleus in the manganese

2. TEMPERATURE DEPENDENCE OF THE NMR FREQUENCY

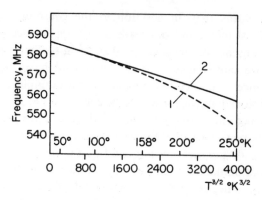

Figure 4

The temperature dependence of the NMR frequency of the Mn^{55} nuclei in $MnFe_2O_4$ (for A sites). Curve 1 is experimental, curve 2 was constructed according to the expression $v = v_0 (1 - 1.2 \cdot 10^{-5} T^{3/2})$ (Heeger and Houston, 1964).

ferrite $MnFe_2O_4$ (Heeger and Houston, 1964). The experimental points for ω_n as a function of $T^{3/2}$ fit well the straight line roughly up to 150°K.

Let us remark, however, that from the theoretical point of view the verification of the "three halves law" at present is no longer of great interest. The point is that $T^{3/2}$ as the first term of an expansion at low temperatures appears practically in all models and theoretical treatments of ferro- and ferrimagnets.* The subsequent terms in the expansion of $\Delta M(T)$ are those which depend significantly on the model and on the character of the theoretical approximations made.

Within the limits of spin wave theory based on the Hamiltonian (2.11), the deviation from the "three halves law", the investigation of which by the NMR method has considerable theoretical interest, may be associated with the following two causes: first, the use of the more accurate expression for the energy of the spin waves (2.21), and, second, the inclusion of the interaction between them.

If, in the expression of the form (2.21a) together with the quadratic terms, also terms of the fourth power, etc., are taken into account when the exchange energy is expanded in powers of k, additional terms $T^{5/2}$, etc., will appear in (2.27):

$$\frac{\Delta M_p(T)}{M_{p0}} = C_{3/2} T^{3/2} + C_{5/2}^{(p)} T^{5/2} + C_{7/2}^{(p)} T^{7/2} + \ldots \qquad (2.28)$$

In the case of ferrimagnets there exists still another reason for the appearance of the term $T^{5/2}$, also associated with the more accurate expression for the energy ε_k (see (2.21)). The point is that for ferrimagnets the coefficient J_{eff} of k^2 in (2.21) contains in fact a small term proportional to the effective field H_p (this has already been men-

---

* This applies also to the theory of ferromagnetism in metals, based on the collective electron model. Together with the term $T^{3/2}$, however, there appears in this theory a term T^2.

tioned before). This means that, in addition to the constant, $\partial \varepsilon_k / \partial H_p$ contains also a term proportional to k^2, and this leads to the appearance of the term $T^{5/2}$ in $M_p(T)$ even in the absence of terms higher than k^2 in ε_k. The coefficients of $T^{5/2}$ associated with the two causes mentioned can, in general, have opposite signs and in some cases they cancel out. Such is the situation, for example, in the manganese ferrite $MnFe_2O_4$ mentioned above, for which the coefficients of $T^{5/2}$ associated with the two causes mentioned cancel out to within 1% (Heeger and Houston, 1964). As a matter of fact, this is the reason why the "three halves law" for the manganese sublattice is satisfied in a very wide temperature interval (see Figure 4).

Let us point out that unlike the coefficient $C_{3/2}$, which in the first approximation (if the magnetic energy is neglected in comparison with the exchange energy) is the same for all the magnetic sublattices of a given ferrimagnet, the coefficients $C_{5/2}^{(p)}$, $C_{7/2}^{(p)}$, etc., can considerably depend on the index p of the sublattice. In general these coefficients are highly sensitive to the actual type of magnetic sublattice structure and to the character of the variation of the exchange energy with distance. It is precisely this circumstance which makes it possible to apply the investigation of deviations from the "three halves law" to the quantitative verification of the conclusions of spin wave theory and to the identification of the nature of the magnetically ordered state of various ferrimagnets.

As was first shown by Dyson (1956), the thermodynamic description of the properties of ferromagnets in the ideal gas approximation of noninteracting spin waves leads to correct results for $\Delta M(T)$ right up to the terms $T^{7/2}$, inclusive. The inclusion of the interaction between spin waves, which is, as has already been mentioned above, the second possible cause for deviation from the "three halves law", leads to additional terms, the largest of which is proportional to T^4. In a number of works it was also shown that the interaction between spin waves can also be taken into account in terms of an ideal gas of spin waves, if its energy, represented by expression (2.21a)*, is renormalized in the proper manner. The renormalization in effect leads to a situation in which some of the parameters of the spin wave energy (for example, J_{eff} in (2.21)) become temperature dependent. In particular, the first temperature correction to J_{eff} is proportional to $T^{5/2}$, and consequently the same correction appears in the coefficient $C_{3/2}$; and this, in the final account, leads to the additional term proportional to $T^{5/2} T^{3/2} \equiv T^4$ in the expansion (2.28).

The compound $CrBr_3$ is a good example of a Heisenberg ferromagnet in which the temperature dependence of the spontaneous magnetization, and of the NMR frequency proportional to it, can be described at low temperatures by an expansion of the form (2.28). In the temperature interval 1–4°K ($T_C \simeq 38$°K) the temperature change of the NMR frequency $\Delta \omega_n(T)$ can be represented by a two-term expression with $T^{3/2}$ and $T^{5/2}$ (Gossard et al., 1961).

---

* See, e.g., Chapter 8 in Mattis (1967).

In ferrimagnetic ferrites with a spinel structure, alongside the manganese ferrite mentioned above, the "three halves law" for the NMR frequency was also observed for the Fe^{57} nuclei in magnetite Fe_3O_4 (Boyd and Slonczewski, 1962). In the two sublattices of lithium ferrite, the spinel Fe_5LiO_8, there is good agreement between the temperature dependence of the spontaneous magnetization obtained from static magnetic measurements and data obtained by way of calculation from the experimental values of NMR frequencies for Fe^{57} (Le Dang Khoi and Bertaut, 1962). Finally, for ferrimagnetic ferrites with a garnet structure ($Y_3Fe_5O_{12}$ and $Lu_3Fe_5O_{12}$), the temperature dependence of the Fe^{57} NMR frequencies in the tetrahedral d- and octahedral a-sublattices can be described by a three-term expression of the form (2.28) with different coefficients $C_{5/2}$ and $C_{7/2}$ for the sublattices d and a (in agreement with theory).

Up to now the discussions centered on the temperature dependence of $\Delta M_p(T)$, which can be obtained from expression (2.25) with condition (2.26). We have thereby completely neglected the effect of the dipole interaction and have mentioned the influence of the field H_{eff} only in connection with the coefficient $C_{5/2}$ in the expression (2.28). In any case, the "three halves law" represented by expression (2.27) is obtained from (2.25) for $\Phi_k = H_{eff} = 0$. The asymptotic expression for $\Delta M_p/M_{p0}$, with condition (2.26), taking into account the first nonvanishing correction arising from the dipole interaction and from H_{eff}, can again be formally represented by an expansion of the form (2.28), the coefficients of which will contain small temperature dependent corrections, H_{eff} and $4\pi M_s$. In particular, the coefficient in front of $T^{3/2}$ will have the form (Holstein and Primakoff, 1940)

$$C_{3/2}\,(1 - cT^{-1/2}),\tag{2.29}$$

where c is a constant, depending on H_{eff} and $4\pi M_s$. For example, if $4\pi M_s \gg H_{eff}$, then $c = 0.53\,(\hbar\gamma_{eff}4\pi M_s/\kappa)^{1/2}$.

Let us point out that the dipole correction presented in expression (2.29) constitutes only the first term in an expansion in powers of the "small parameter" $\hbar\gamma_{eff}4\pi M_s/\kappa T$. Since the parameter appears here raised to the power 1/2, the dipole correction is more important than could have been assumed beforehand. (Thus, for example, for iron at 10°K the second term in expression (2.29) is already about 1/4.) Therefore, in a number of ferro- and ferrimagnets, the temperature interval in which the "three halves law" could have been realized in a pure form may in general be missing. In particular, this refers to substances with a low Curie temperature and a high value of the saturation magnetization M_s. An example of a ferromagnet of this type is the compound EuS ($T_C \simeq 16°K$ and $4\pi M_s \simeq 14\,kOe$) for which the expansion presented above (2.28) is in general useless (even if a dipole correction of the form (2.29) is taken into account). Charap and Boyd (1964) described the experimental data which they had obtained with the use of the NMR method, on the basis of the general expression (2.25), using for the spin wave energy expression (2.20) in which (2.21a) was taken

as ε_k, and the exchange interaction was taken into account inside two coordination spheres. (Naturally, this calculation could be carried out only with the use of a computer.) As a result of such a treatment it proved possible to reconcile the conclusions of the spin wave theory for the spontaneous magnetization and for the magnetic part of heat capacity.

Antiferromagnets

Let us turn to the examination of antiferromagnets. The problem of the verification of the NMR method of spin wave theory for antiferromagnets is discussed in detail in Review (1965a). For this reason we shall dwell on it very briefly.

It can generally be shown that for antiferromagnets with collinear magnetic structures, there are two branches of acoustic spin waves, while the remaining $n - 2$ branches (if $n > 2$) are optical (Turov, 1963), irrespective of the number of sublattices n. However, the acoustic branches of the spectrum have important distinctive properties in antiferromagnets in comparison with ferro- and ferrimagnets. We shall illustrate this point by an example of an antiferromagnet with two equivalent magnetic sublattices for which $\gamma_1 = \gamma_2 = \gamma_e$, $S_1 = S_2 = S$, $H_1 = H + H_a$ and $H_2 = H - H_a$, and also identical exchange integrals between the corresponding spin parameters of both sublattices. The dipole interactions are usually neglected (because of the absence of a macroscopic magnetic moment in the ground state of the antiferromagnet at $H = 0$).

The spin wave energy of such an antiferromagnet for both branches of the spectrum can be described by a single expression in the form (Bogolyubov and Tyablikov, 1949):

$$\varepsilon_k^{(1,2)} = [(\alpha_1(k) + \alpha_2(k))^2 - \beta^2(k)]^{1/2} \pm \tfrac{1}{2}\gamma_e(H_1 + H_2), \qquad (2.30)$$

$$\alpha_{1,2} = 2S[J(0) - J(k) + J_{12}(0)] \pm \tfrac{1}{2}\gamma_e\hbar H_{1,2}, \qquad (2.31)$$

$$\beta(k) = 4SJ_{12}(k). \qquad (2.32)$$

Here

$$J(k) = \sum_R J_{11}(R)e^{-ikR} = \sum_R J_{22}(R)e^{-ikR} \qquad (2.33)$$

and

$$J_{12}(k) = -\sum_R J_{12}(R)e^{-ikR} \qquad (2.34)$$

constitute the Fourier components of the exchange integrals inside each sublattice and between them respectively. The "minus" sign in expression (2.34) is introduced in connection with a more convenient notation taking into account the fact that for an antiferromagnet $J_{12} < 0$.

The peculiar spectral features consist in the following. First, even for the acoustic branches of the spectrum (see expression (2.30)) the energy gap depends on the exchange interaction. Indeed for $k = 0$ we have from (2.30)

$$\varepsilon_0^{(1,2)} = |\gamma_e|\hbar\left(\sqrt{(H_E + H_a)H_a} \mp H\right),\qquad(2.35)$$

where $H_E = 8SJ_{12}(0)/|\gamma_e\hbar|$ is the effective field of the exchange interaction between the sublattices. Since usually $H_E \gg H_a$ ($H_E \sim 10^6$–10^7 Oe and $H_a \sim 10^2$–10^4 Oe, the energy gap is determined by the effective field

$$H_{Ea} \simeq \sqrt{H_E H_a},\qquad(2.36)$$

which constitutes the geometric mean value of the anisotropy and exchange fields and attains a magnitude of 10^4–10^5 Oe. This means that for antiferromagnets the energy gap for spin waves

$$\varepsilon_0 \sim \kappa T_{Ea} = \hbar\omega_{Ea}\qquad(2.37)$$

may attain a few or even tens of degrees Kelvin ($T_{Ea} = h\,|\gamma_e|\,H_{Ea}/\kappa$), while in frequency units ($\omega_{Ea} = |\gamma_e|H_{Ea}$) the gap may be of the order of 10^{12} sec^{-1} and more.*

The second peculiarity of antiferromagnets is that the spin wave energy for sufficiently large k, such that $\varepsilon_k \gg \varepsilon_0$ (and at the same time sufficiently small so that $J(k)$ and $J_{12}(k)$ can be expanded in powers of k), is not a quadratic but a linear function of k. Indeed, the spin wave energy defined by expression (2.30) can be represented up to a quadratic term in k for $H = 0$ in the form

$$\varepsilon_k = \varepsilon_k^{(1)} = \varepsilon_k^{(2)} = \sqrt{J^2(ak)^2 + \varepsilon_0^2},\qquad(2.38)$$

where J is (together with H_E) the second parameter of the exchange interaction, defined generally by the relation**

$$J^2(ak)^2 \simeq (4S)^2\left[(J(0) - J(k) + J_{12}(0))^2 - J_{12}^2(k)\right].\qquad(2.39)$$

Thus from (2.38) we have, for $J(ak) \gg \varepsilon_0$

$$\varepsilon_k \simeq J(ak).\qquad(2.40)$$

Expression (2.19) for the temperature dependence of the sublattice magnetization (for $H = 0$) takes, when applied to an antiferromagnet, the following form

$$\Delta M(T) = |\gamma_e|\hbar\sum_k \varepsilon_k^{-1}\bar{n}_k\{4S[J_{12}(0) + J(0) - J(k)] + \hbar\gamma_e H_a\}.\qquad(2.41)$$

---

* In this case, the name acoustic for these branches of the spectrum is even more arbitrary than for ferromagnets.

** In particular, for a simple cubic lattice in the approximation of nearest neighbors, when these neighbors are atoms of another sublattice, the parameter J is connected with H_E in the following manner:

$$J = \hbar|\gamma_e|H_E/\sqrt{12}.$$

Here $\varepsilon_k = \varepsilon_{1k} = \varepsilon_{2k}$ is the spin wave energy, defined by the expression

$$\varepsilon_k = \varepsilon_{1k} = \varepsilon_{2k} = \{[4S(J_{12}(0) + J(0) - J(k)) + |\gamma_e|\hbar H_a]^2 - (4SJ_{12}(0))^2\}^{1/2}, \quad (2.42)$$

and $\bar{n}_{1k} = \bar{n}_{2k} = \bar{n}_k$ is the spin wave number identical at $H = 0$ for both branches of the spectrum*

In the temperature range

$$T_{Ea} \ll T \ll T_N \tag{2.43}$$

the approximate expression (2.40) can be used for ε_k, and in this case the first term of the expansion has the form

$$\frac{\Delta M(T)}{M_0} = C_2 T^2. \tag{2.44}$$

Relation (2.44) replaces the "three halves law" for an antiferromagnet.

In many antiferromagnets, however, the temperature interval indicated in (2.43) is either altogether absent or very small. For example, for the best studied (by NMR methods) antiferromagnetic compound MnF_2, $T_{Ea} \simeq 13°K$ and $T_N \simeq 67°K$. In this case, relation (2.41) can be used for the calculation of $M(T)$ and the more general expression (2.38) for ε_k. But usually even the applicability of this expression is very limited. This is so because for temperatures $T \gtrsim T_{Ea}$ spin waves can be excited with such large k that the expansion of $J_{12}(k)$ and $J(k)$ in powers of k up to quadratic terms becomes no longer applicable.** The existence of Brillouin zone boundaries in k space becomes evident. Nevertheless, the theory of free spin waves, within the framework of which expression (2.41) was obtained, can still be applied, provided the expansion mentioned is carried to much higher powers of k (or, in general, the exact nonperiodic functions (2.33) and (2.34) are used for $J_{12}(k)$ and $J(k)$ in the calculation of $M(T)$, limiting the summation on R inside a small number, one or two, of coordination spheres). Such calculations are again carried out by numerical

---

* Let us point out that for crystals of rhombic or lower symmetry the equality of the spin wave energy (and hence of the number of spin waves) for both branches of the spectrum cannot in general be confirmed (Turov, 1963, page 86).

** Here it is appropriate to turn our attention to a certain arbitrary feature of expression (2.21) for a ferromagnet, or expression (2.38) for an antiferromagnet, namely that in these expressions only the dependence of ε_k on the magnitude of the vector k is shown. Actually, for crystals the symmetry of which is lower than cubic, the quadratic terms of the exchange energy depend also on the direction of k. This anisotropy of the spin wave dispersion relation (i.e., the dependence of the energy on k), due to the difference in the exchange interaction for different directions in the crystal, can be very substantial in the case of a so-called layer structure, examples of which are $FeCl_2$ and $CrCl_3$. For these compounds there are strong exchange (ferromagnetic) interactions inside the layer and weak antiferromagnetic interactions between the layers. In some temperature intervals such magnets behave like two-dimensional ferromagnetic lattices for which theory predicts a linear temperature dependence, instead of (2.27) or (2.44). Narath (1963) observed such a dependence of the NMR frequency of Cr^{53} in $CrCl_3$ (see also Review, 1967).

methods. For the antiferromagnet MnF_2 mentioned above, the temperature dependence of the NMR frequency (for F^{19} nuclei) can, right up to the temperature $T \simeq (1/3)T_N$, be described within the framework of the rigorous theory of free spin waves, provided the exchange interaction and anisotropy parameters are appropriately selected. Note that in the quadratic, with respect to k, approximation (2.38) of the energy ε_k, this limiting temperature is lowered to 0.1 T_N (Review, 1965a).

All this refers mainly to the so-called "easy axis" type (EA) antiferromagnets, for which the magnetic moments are ordered along some definite crystallographic direction (for example, along the main symmetry axis of a uniaxial crystal or one of the higher axes of symmetry of a cubic crystal). The important feature of these antiferromagnets consists of the fact that in them the deviation of the antiferromagnetic axis (characterized by the vector $L = M_1 - M_2$) from its "easy direction" requires a very large magnetic field $H \gtrsim H_{Ea} = \sqrt{H_E H_a}$.*

There exists another type of antiferromagnets for which the antiferromagnetic axis lies in a plane with very small magnetic anisotropy (for example, in the basal plane of a hexagonal crystal). For such antiferromagnets (of the "easy plane" type, EP) even a small external field applied in this plane will orient the antiferromagnetic axis perpendicular to itself. The main difference between the two types is that for EP-type antiferromagnets at least one of the branches of the spin wave spectrum has a small energy gap, determinable in real experimental conditions mainly by the external magnetic field. As has already been mentioned in Chapter I, in the presence of such a "low-frequency" branch of spin waves a dynamic coupling between the vibration of the electron and nuclear spins may essentially appear, which can lead to an additional rather large and temperature-dependent shift in the nuclear and electron resonance frequencies.* This case will be dealt with in the next section.

Concluding the review on the results of the verification of the conclusions of spin wave theory relating to the temperature dependence of the NMR frequencies, it can be stated that these conclusions are basically confirmed. This is certainly true for nonmetallic ferro-, ferri- and antiferromagnets. (For metals, in our view, there are no sufficiently reliable investigations of this problem yet.) In the future it will be necessary to pay particular attention to the quantitative side of the matter. We have in mind the quantitative comparison of the parameters of the theory (exchange integrals, anisotropy constant, etc.) obtained by other methods, as well as NMR: from static magnetic measurements, measurements of ferro- and antiferromagnetic resonance and spin wave resonance, the magnetic part of heat capacity, etc.

---

* The latter emerges from expression (2.35) for the magnitude of the energy gaps: one of them becomes negative for $H \gtrsim \sqrt{H_E H_a}$. This shows that the situation assumed by us, in which the antiferromagnetic axis coincides with the "easy axis," becomes unstable only for such fields.

** Indeed, this subdivision of antiferromagnets into two types (EA and EP) is quite arbitrary. It loses all meaning for cubic antiferromagnets with sufficiently small magnetic anisotropy, for which, in view of the smallness of H_a, both branches of spin waves are of the "low-frequency" type.

The Neighborhood of the Curie (Néel) Point

Today one of the most fundamental problems of theoretical physics is the construction of a consistent theory of phase transitions of the second kind. The precise measurement of the temperature dependence of the sublattice magnetization near the magnetic transition point Θ (T_C or T_N) has great significance for the development of this theory.

The thermodynamic theory of Landau (1937) and the molecular field theory of Weiss–Néel and of Van Vleck* predict for $M(T)$ near the point Θ (below it) an expression of the form

$$M(T) = \text{const}\,(\Theta - T)^\beta, \tag{2.45}$$

where $\beta = 1/2$. The first of these theories based on the expansion of the thermodynamic potential of the magnetic crystal in a power series in M near the point Θ. This is done using the important (though in no way justified) assumption that the coefficients of such an expansion are analytic functions of the temperature and of the pressure. In the molecular field theory, the energy spectrum of the spin system (which is unknown) is replaced in the calculation of the thermodynamic potential by some mean energy value ("the energy center of gravity") depending only on the magnetization of the sublattices M_p.

Meanwhile the exact solution of the problem of ferromagnetic transition, given by Onsager (1944) for the case of the two-dimensional Ising model, showed that the thermodynamic potential of the spin system at the point Θ has a nonanalytical singularity, while the power index β for the magnetization defined by the relation (2.45) is equal to 1/8 (see, for example, the book of Mattis, 1967). Unfortunately, an analogical exact solution for the three-dimensional problem is not yet available even for the Ising model. However, various approximate methods as well as precise experimental measurements of other thermodynamic quantities (for example, heat capacity and magnetic susceptibility) near a phase transition of the second kind have confirmed the basic conclusion of Onsager on the presence of singularities in the thermodynamic potential. At the same time the index β in (2.45) was shown to be considerably dependent on the model (and, in particular, on the number of dimensions). The so-called Pade approximation method** for the three-dimensional model yielded $\beta = 5/16$.

Only a few years ago it was thought that the result of the Landau theory and of the molecular field theory for $M(T)$, represented by expression (2.45) with $\beta = 1/2$, was confirmed experimentally (see, for example, the book of Belov, 1959).

---

* For a description of these works, see, for example, the books of Vonsovskii (1953) or Borovik-Romanov (1962).

** A numerical method in which a large, but finite number of terms of the series are approximated by an expression of the form (2.45) with a suitable value of the index β. See, in this connection, for example, the work of Gammel et al. (1963).

But in 1962 Heller and Benedek measured the temperature dependence of the NMR frequency of F^{19} in the antiferromagnetic MnF_2 ($T_N \simeq 67°K$) and reached the unexpected conclusion that, in the temperature interval $0.005°K < T - T_N < 2°K$, the experimental data could be well described by (2.45) with $\beta = 1/3$ (more accurately, $\beta = 0.333 \pm 0.007$). This is illustrated in Figure 5, taken from this work. Later, Heller and Benedek (1965a) obtained the analogous result for the ferromagnetic compound EuS ($T_C \simeq 16.5°K$). The temperature dependence of the resonance of the Eu^{151} and Eu^{153} nuclei in the temperature interval $0.2°K \leq T_C - T < 2°K$ could again be described by expression (2.45) with $\beta = 0.33 \pm 0.015$.

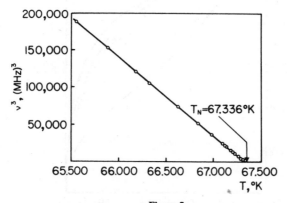

Figure 5

The temperature dependence of the cube of the NMR frequency of F^{19} in MnF_2 in an interval of $1.8°$ below the Néel temperature (Heller, 1966).

The experimental value $\beta \simeq 1/3$ is very close to the theoretical value mentioned above for the three-dimensional Ising method ($\beta = 5/16$).

A more detailed discussion of this problem may be found in the Proceedings of of the International Conference on Magnetism (Nottingham, 1964) in the reports of Fisher (1965), Heller and Benedek (1965b) and also in the works of Van Loef (1966), Senturia and Benedek (1966), Eibschitz et al. (1966).

Impurity Atoms

Measurements of the temperature dependence of the NMR frequency for the nuclei of impurity atoms are of considerable interest in the construction of consistent models of ferromagnets and antiferromagnets. The measurements (side by side with the Mössbauer effect) allow the investigation of the temperature dependence of local magnetic fields for various combinations of matrix and impurity. Such information is necessary not only for the evaluation of the exchange interactions between various magnetic atoms of an alloy or a compound (within the framework

of the localized spin model) but it also allows the detection of a possible redistribution with the temperature of the spin density near the impurity nucleus.

In the work of Koi et al. (1964), the temperature dependence of the frequency ω_n for the Mn^{55} nuclei in a dilute (1.5% Mn) ferromagnetic Fe–Mn alloy (Figure 6) was investigated. The latter turned out to differ substantially from the temperature dependence of the magnetization $M(T)$ of the matrix (iron). In particular, $\omega_n(T)$ decreased with temperature considerably more rapidly than $M(T)$.

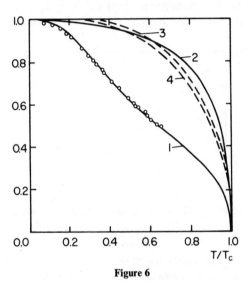

Figure 6

The temperature dependence of the NMR frequency of the Mn^{55} nuclei in iron.

Curve 1 was calculated from expressions (2.46)–(2.48); the experimental data are shown on it by circles. Curve 2 represents the experimental temperature dependence of the magnetization of the matrix (iron). It can be compared with the theoretical curves 3 and 4, calculated from the molecular field theory for $S = 1/2$ and $S = 1$ respectively.

Jaccarino et al. (1964) have explained this experimental fact on the basis of the following assumptions: 1) The frequency $\omega_n(T)$ is proportional to the thermal mean value of the manganese spin $\sigma_{Mn}(T)$; 2) The manganese spin S_{Mn} is localized and its magnitude $|S_{Mn}|$ does not depend on temperature; 3) The magnetic energy levels of the manganese atom are discrete quasi-Zeeman levels of its spin in the molecular exchange field $H_{Mn}(T)$, proportional to the magnetization of the matrix:

$$H_{Mn}(T) = H_{Mn}(0) \frac{\sigma_{Fe}(T)}{\sigma_{Fe}(0)}. \tag{2.46}$$

On the basis of these assumptions the temperature dependence of the Mn^{55} NMR frequency must be described by the Brillouin function (see, for example, the

book of Kittel, 1963, p. 246):

$$\frac{\omega_n(T)}{\omega_n(0)} = \frac{\sigma_{Mn}(T)}{\sigma_{Mn}(0)} = B_{S_{Mn}}(x),$$ (2.47)

where

$$x = \frac{\mu_B\, g_{Mn} S_{Mn}}{\kappa T}\, H_{Mn}(T).$$ (2.48)

And indeed, for values $H_{Mn}(0) = 3.7 \cdot 10^6$ Oe, $S_{Mn} = 3/2$ and $g_{Mn} = 2.00$ of the parameters, the experimental points fit very well the curve defined by expressions (2.46)–(2.48).

The value of the exchange field $H_{Mn}(0)$ thus obtained, which acts on the impurity spins (manganese) due to the matrix (iron) spins, can now be compared with the exchange field $H_{Fe}(0)$, characteristic of the interaction between the spins of the matrix. The latter, in accordance with the Weiss molecular field theory (see again Kittel, 1963), can be expressed through the Curie temperature T_C in the following way:

$$H_{Fe}(0) = \frac{3\kappa T_C}{\mu_B g_{Fe}(S_{Fe} + 1)}.$$

Assuming for iron $T_C \simeq 1000°K$, $S_{Fe} = 1$ and $g_{Fe} = 2$, we have $H_{Fe}(0) \simeq 11 \cdot 10^6$ Oe. Consequently, $H_{Mn}(0)/H_{Fe}(0) \simeq 1/3$.

The essential difference between the temperature dependence of σ_{Mn} (and, consequently, of the Mn[55] NMR frequency) and that of σ_{Fe}, obtained by such a treatment of the experimental data, is associated with the fact that the exchange interaction between the impurity atom and the surrounding atoms of the matrix is considerably less than the exchange interaction between the matrix atoms themselves.

It is clear that, being based on the above-mentioned rough assumptions (particularly in application to a metal), expression (2.47) can be considered only as a semiempirical description of the phenomenon. Nonetheless it is a very interesting fact that the molecular field theory is by far more accurate when applied to the impurity than when applied to the matrix. The latter is clearly seen in Figure 6 (Callen et al., 1965) in which side by side with the experimental curve $\sigma_{Fe}(T)/\sigma_{Fe}(0)$ also the theoretical curves calculated according to the Weiss theory for $S = 1$ and $S = 1/2$ are presented. This situation requires a more detailed theoretical consideration, which is indeed done in a number of works*. It has been shown by the use of the Green's function method that a more rigorous theory of the magnetic properties of a crystal containing impurity atoms can also, in a certain particular case, lead to an expression of the form (2.47) for the temperature dependence of the mean impurity spin.

* Callen et al. (1965), Izyumov and Medvedev (1966), Wolfram and Hall (1966), Hone et al. (1966).

Such a particular case is exactly the case of a weakly bound impurity, for which the exchange interaction between the impurity and the matrix is considerably less than between the atoms of the matrix. In the quasicontinuous spectrum of the elementary excitations of such a crystal there are narrow so-called resonance levels E_s, situated near the bottom of the spin-wave band of the matrix. To this state there corresponds a sharp peak in the density of states curve, and the amplitude of the spin deflections in this state has a sharp maximum at the impurity atom. When the temperature is increased the level E_s situated near the bottom of the band begins to fill up fast and this causes a sharp decline in the mean magnetic moment of the impurity atom.

This theory in the first, very rough approximation leads to the following expression for the mean spin $\sigma' \equiv \langle S'_z \rangle$ of the impurity atom (Izyumov and Medvedev, 1966):

$$\sigma'(T) \equiv S'B_{S'}\left(\frac{S'E_s}{\kappa T}\right), \tag{2.49}$$

where

$$E_s \simeq 2\sigma Z J' \tag{2.50}$$

is the energy defining the location of the resonance level at the given temperature; $\sigma \equiv \langle S_z \rangle$ is the mean spin of the matrix atom, J' is the parameter of the exchange coupling between the impurity and the matrix and Z is the number of nearest neighbors ($B_{S'}$ is again the Brillouin function). Expressions (2.47) and (2.49) will be identical if we assume $g_{Mn}\mu_B H_{Mn} = E_s$, and this, taking (2.50) into account, in fact means that the molecular field $H_{Mn}(T)$ has the very same form as that which follows from the Frenkel–Heisenberg theory of ferromagnetism (Kittel, 1963).

It should be noted that the theoretical treatment described above refers to a model of a ferromagnet based on the concepts of localized spins which is applicable, strictly speaking, only to ferrodielectrics. It is possible that for this reason the previous discussion which explains the experimental results of NMR in the Fe–Mn alloy, leads, at the same time, also to a difficulty. The latter stems from the fact that the value of the Mn atomic spin ($S = 3/2$) obtained by the above-mentioned method, is inconsistent with the neutron diffraction studies of this alloy. These experiments allegedly show that the magnetic moment of the Mn atom does not exceed $\frac{1}{2}\mu_B$ (Collins and Low, 1965). In the work of Low (1966) which resolved this contradiction it is suggested that in addition the contribution of localized spins (described by an expression of the form (2.47)) to the hyperfine field, the field set up at the nucleus by the collective electrons should also be taken into account. It is moreover assumed that the latter, as distinct from the former, is proportional not to the mean impurity spin $\sigma_{Mn}(T)$, but to the mean matrix spin $\sigma_{Fe}(T)$. As a result it is possible to describe the experimental data without having to resort to such high values for the manganese spin, simply by introducing another parameter.

Another example of an alloy in which a different temperature dependence was found for the mean spin of the impurity (Fe) and of the matrix (Ni) is the alloy Ni–Fe (Dash et al., 1966). In this work, however, the Mössbauer effect and not NMR was investigated, and we shall mention it only in connection with the fact that the iron atoms in nickel, as distinct from manganese in iron, constitute an example of a strongly coupled impurity. The authors discuss and compare both these cases. On NMR in alloys see also the interesting article of Jaccarino (1968).

3. Coupled Electron and Nuclear Spin Vibrations

We have up to now considered the HFI only as a cause of the static local field acting on the nuclear spin and leading to a static shift in the NMR resonance frequency, equal to $\omega_{n0} = A\sigma/\hbar$. In fact, the HFI can also give rise to a coupling between the electron and the nuclear subsystem spin vibrations, which leads to an additional dynamic shift in the resonance frequencies of both subsystems (De Gennes et al., 1963). In some cases (mainly for antiferromagnets of the "easy plane" type) the dynamic shift mentioned can be considerable, and, since it depends on the temperature, it is, in these cases, necessary to take it into account in the investigation by the NMR method of the temperature dependence of magnetization.

Ferromagnets

Let us consider first the very simple case of a ferromagnet magnetized to saturation, which can be described by a single uniform electron magnetization

$$M = \gamma_e \hbar N \langle S \rangle$$

(a single "magnetic sublattice") and uniform nuclear magnetization

$$m = \gamma_n \hbar N \langle I \rangle .$$

We shall take into account only the isotropic HFI of the form (1.4) between the electron and the nuclear subsystem spins. The total energy density of a ferromagnet in an external magnetic field $H_z = H$, applied along the axis of easy magnetization, can then be described in the following form*

$$\mathscr{H} = \mathscr{H}_0 + A_0 M m - M_z H_a - (M_z + m_z)H . \tag{2.51}$$

Here $\mathscr{H}_0$ is the part of the energy which does not depend on the orientation of the vectors M and m, H_a is the effective field of the magnetic anisotropy.

---

* Strictly speaking, in the case of finite temperatures, (2.51) represents not the energy but the thermodynamic potential of the spin system.

So as not to complicate matters by taking into account the shape anisotropy, we have assumed that the specimen has a spherical form. Besides, we have here completely ignored the dipole field at the nuclei, which, strictly speaking, is valid only for nuclei the surroundings of which have cubic symmetry.

The problem of finding the uniform eigenvibrations of two coupled magnetic moments M and m with an energy defined by expression (2.51) is completely analogous in form to the corresponding problem for the vibrations of the magnetic moments M_1 and M_2 of a ferrite with two magnetic sublattices (see, for example, the book of Gurevich, 1960). According to classical concepts, which we shall adopt here, it is necessary to start with the system of equations

$$\dot{M} = \gamma_e [M \times H_M],$$
$$\dot{m} = \gamma_n [m \times H_m], \tag{2.52}$$

where

$$H_M = - \frac{\partial \mathscr{H}}{\partial M}$$

and

$$H_m = - \frac{\partial \mathscr{H}}{\partial m}.$$

In the equilibrium state $M_z = M$,

$$m_z \equiv m = \chi_n(H + H_n)$$

and

$$M_x = M_y = m_x = m_y = 0.$$

For small vibrations near the equilibrium, equation (2.52) can be linearized with respect to the transverse components of M and m while the longitudinal components remain, in the first approximation, equal to their equilibrium value: $M_z \simeq M$ and $m_z \simeq m$. If we introduce cyclic variables

$$M^\pm = M_x \pm iM_y \text{ and } m^\pm = m_x \pm im_y,$$

equations (2.52) are reduced, in the linear approximation, to the form

$$\dot{M}^\pm = \pm i\gamma_e [- M^\pm(- A_0 m + H + H_A) - m^\pm A_0 M], \tag{2.53}$$

$$\dot{m}^\pm = \pm i\gamma_n [- m^\pm(- A_0 M + H) - M^\pm A_0 m]. \tag{2.53a}$$

Assuming

$$M^\pm = M_0^\pm \exp(i\omega t) \quad \text{and} \quad m^\pm = m_0^\pm \exp(i\omega t),$$

we obtain, by equating to zero the determinant of the system of equations for $M_0^{\pm}$ and $m_0^{\pm}$, two equations which define the eigenfrequencies

$$[\pm\omega + \gamma_n(- A_0 M + H)][\pm\omega + \gamma_e(- A_0 m + H_a)] - \gamma_e\gamma_n A_0^2 Mm = 0. \quad (2.54)$$

The four roots of these two equations form two pairs which differ in sign only and therefore there are two positive roots having a physical meaning,

$$\omega_{1,2} = |\tfrac{1}{2}[\gamma_e(- A_0 m + H + H_a) + \gamma_n(- A_0 M + H)] \pm$$

$$\pm \{\tfrac{1}{4}[\gamma_e(- A_0 m + H + H_a) - \gamma_n(- A_0 M + H)]^2 + \gamma_n\gamma_e A_0 Mm\}^{1/2}|. \quad (2.55)$$

In the zero order approximation in nuclear magnetization ($m \to 0$), these two roots become the unperturbed electron resonance frequency (the FMR frequency)

$$\omega_1 \simeq \omega_e = |\gamma_e|(H + H_a) \quad (2.56)$$

and the nuclear resonance frequency

$$\omega_2 \simeq \omega_n = |\gamma_n(H_n + H)|, \quad (2.57)$$

respectively.

In the next approximation in m, the frequencies ω_1 and ω_2 are shifted proportionally to m

$$\omega_1 = \omega_e\left(1 + \eta\frac{m}{M}\right) \equiv \Omega_e, \quad (2.58)$$

$$\omega_2 = \gamma_n\left|H_n\left(1 - \eta\frac{m}{M}\right) + H\right| \equiv \Omega_n, \quad (2.59)$$

where $\eta = H_n/(H + H_a)$ is the amplification coefficient introduced earlier (see expression (1.9)); to fix ideas, we have assumed that $\gamma_n > 0$.

The nuclear magnetization m depends considerably on the temperature ($m \sim 1/T$), and, thus, the presence of a mutual coupling between the nuclear and the electron spin vibrations leads, with the lowering of the temperature, to a temperature shift in the nuclear and electron resonance frequencies. The relative values of these frequency shifts are equal in magnitude but opposite in sign:

$$-\frac{\delta\omega_n}{\omega_n} = \frac{\delta\omega_e}{\omega_e} = \eta\frac{m}{M} = \frac{I(I + 1)}{3S}\frac{\omega_n}{\omega_e}\frac{\hbar\omega_n}{\kappa T}. \quad (2.60)$$

Assuming, for instance, $\omega_n = 3 \cdot 10^9$ sec^{-1}, $H_a = 10^3$ Oe and $I = S = 5/2$ (these values of I and S correspond to the Mn^{++} ion) we obtain

$$\frac{\delta\omega}{\omega} \sim \frac{3 \cdot 10^{-3}}{T}. \quad (2.61)$$

For NMR this shift can be noticeable only at temperatures of the order of or lower than $1°$K (where it becomes equal to the line width). It can be expected that this shift

will be larger for nuclear resonance in the domain walls, where $\eta \gtrsim 10^3$ (see Chapter I).

Let us further remark that the change of $m \equiv m_z$ and the shift in resonance frequency associated with it can take place not only on account of the change in the temperature of the lattice but also on account of NMR saturation. At sufficiently large intensities, the longitudinal magnetization, m_z, and consequently also the resonance frequencies (2.58) and (2.59) will depend, owing to nonlinearity, on the amplitude of the radio-frequency field. The condition of NMR saturation has the usual form (Abragam, 1963),

$$(\gamma_n h_x \eta)^2 T_1 T_2 \gtrsim 1 , \tag{2.62}$$

but with the difference that the amplitude of the radio-frequency field h_x enters with an amplification factor η (Portis and Gossard, 1960). Here T_1 and T_2 are the longitudinal and transverse relaxation times, respectively, for the nuclear spin-system. Expression (2.60) will be applicable also to the shift in the resonance frequency caused by the NMR saturation effect, provided the lattice temperature T is replaced in it by the spin temperature T_n of the nuclear spin-system.

The dependence, through m_z, of the FMR frequency on the level of the r.f. field intensity at the NMR frequency allows, in principle, the observation of interesting phenomena by double resonance. The effects are, however, much more pronounced in antiferromagnets and will be considered below.

Finally, it is necessary to bear in mind that the shift in the NMR and FMR frequencies due to electron and nuclear spin vibrations depends, in general, on the character of the magnetic anisotropy of the ferromagnet. Both the crystallographic anisotropy and the shape anisotropy of the sample are important. The expressions discussed above are suitable for the case of a uniaxial or a cubic ferromagnet magnetized along the axis of easy magnetization. A more detailed investigation of the problem of coupled vibrations in an electron–nuclear spin system in ferromagnets for various cases can be found in the works of Onoprienko (1965) and Ignatsenko and Kudenko (1966).

Antiferromagnets

As usual, we will describe an antiferromagnet by two magnetizations M_1 and M_2 in accordance with the simple model of two magnetic sublattices. The nuclear spins too will be divided into two sublattices, since the local hyperfine fields at the nuclei at the sites of different sublattices will have opposite signs. We will describe the nuclear spin-system by the magnetizations m_1 and m_2. The discussion will center on the example of crystals with axial symmetry.

The energy (more precisely, the thermodynamic potential) of the whole electron–nuclear spin-system can, in this case, be represented in the form (Turov and Kuleyev, 1965)

$$\mathcal{H} = \mathcal{H}_0 + J_0 M_1 M_2 + \tfrac{1}{2} a (M_{1z} + M_{2z})^2 + \tfrac{1}{2} b (M_{1z} - M_{2z})^2 +$$
$$+ A_0 (M_1 m_1 + M_2 m_2) - (M_1 + M_2 + m_1 + m_2) H + \tfrac{1}{2} (\chi_n/2)^{-1} (m_1^2 + m_2^2). \tag{2.63}$$

Here the second term with the constant $J_0 > 0$ is the exchange energy, while the third and fourth terms represent the anisotropy energy of the antiferromagnetic system; the following two terms are the hyperfine interaction energy and the energy in the external field. The last term in (2.63) could have been included in the constant $\mathcal{H}_0$, which does not depend on the direction of the vectors M_p and m_p ($p = 1, 2$). This term constitutes that part of the thermodynamic potential which depends on the magnitude of the magnetizations m_1 and m_2 and which we need only for the determination of the equilibrium values of m_1 and m_2. The main symmetry axis of the crystal is taken by us as the Z axis of the coordinate system.

In uniaxial crystals there may exist two antiferromagnetic states: a state in which the antiferromagnetic axis (characterized by the equilibrium direction of the vector $L = M_1 - M_2$) coincides with the main symmetry axis of the crystal, and a state in which the axes mentioned are perpendicular to one another, so that the vector L lies in the basal plane. The first of these states ($L \| Z$) is realized in the case when the anisotropy constant $b < 0$, while the second ($L \perp Z$) when $b > 0$.

We have already mentioned in a previous section that the magnitudes of both eigenfrequencies of the vibrations of the magnetic moments M_1 and M_2 in the state with $L \| Z$ (the state of EA type) are of the order of

$$\omega_{Ea} = |\gamma_e| \sqrt{H_E H_a}, \tag{2.64}$$

where $H_E = 2 J_0 M_0$ is the exchange field* of 10^5–10^7 Oe while $H_a = 2b M_0$ is the magnetic anisotropy field of the order of 10^2–10^4 Oe. Here M_0 is the modulus of the magnetization vector of a single sublattice: $M_1^2 = M_2^2 = M_0^2$. The frequency ω_{Ea} is always considerably higher (by two–three orders) than the nuclear spin precession frequency in the hyperfine field $\omega_{n0} = \gamma_n H_n$. Therefore the "mixing" of the electron and nuclear spin vibrations constitutes in this case a very small effect.

At the same time for the state $L \perp Z$ (the state of EP type) one of the branches of the spectrum of the antiferromagnetic system (corresponding to the vibrations of M_1 and M_2 in the basal plane) has a considerably smaller vibration frequency due to the fact that in the basal plane the anisotropy can be very small. As we shall see below, the mutual coupling between the vibrations of the nuclear spins and this vibration branch of the electron system can be very substantial.

For the reason mentioned we shall carry out a detailed examination of the spectrum

---

* Let us point out that the exchange field H_E introduced by us is twice as large as the corresponding magnitude which usually appears in the literature. In our notation, the factor 2, which in the old notation is always present, disappears.

of the electron–nuclear spin-system eigenfrequencies only for the case of an antiferromagnet of the EP type.*

Let also the constant external field H lie in the basal plane. If the X axis is oriented along H, the resulting equilibrium magnetization of the electron system $M = M_1 + M_2$ will be directed along the same axis, and the vector L will point along the Y axis. The independent variables which determine the equilibrium state can be chosen as follows (Figure 7):

$$M_{1x} = M_{2x} = \tfrac{1}{2} M_\perp , \quad m_{1x} = m_{2x} = \tfrac{1}{2} m_\perp , \quad m_{1y} = - m_{2y} = m_0 . \qquad (2.65)$$

The components $M_{1y} = - M_{2y}$ are not independent variables, since for temperatures sufficiently removed from the Néel point, magnetization occurs by rotation of the vectors M_1 and M_2 without a change in their lengths. Consequently, $M_{py}^2 = M_0^2 - (M_\perp/2)^2$ $(p = 1,2)$. From the conditions

$$\frac{\partial \mathcal{H}}{\partial m_\perp} = 0 , \quad \frac{\partial \mathcal{H}}{\partial m_0} = 0$$

and

$$\frac{\partial \mathcal{H}}{\partial M_\perp} = 0$$

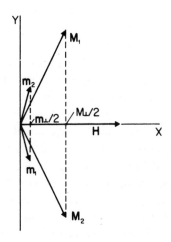

Figure 7

The equilibrium distribution of the electron (M_1 and M_2) and nuclear (m_1 and m_2) magnetizations in an antiferromagnet of the EP type.

---

* See, for example, Lee et al. (1963), Heeger and Teaney (1964), Fink and Shaltiel (1964), Turov and Kuleyev (1965).

we find approximately

$$M_\perp = \frac{H}{J_0}(1 - \tfrac{1}{2}A_0\chi_n) \simeq \frac{H}{J_0} \equiv 2M_0\frac{H}{H_E} = \chi_\perp H \,,$$

(2.66)

$$m_\perp \simeq \chi_n \left(1 - \frac{A_0}{2J_0}\right) H, \quad m_0 = -\tfrac{1}{2}\chi_n A_0 M_0 = \tfrac{1}{2}\chi_n H_n.$$

($\chi_\perp$ and χ_n are the static susceptibilities of the electron and nuclear spin-systems).

For the calculation of the vibration frequency spectrum of the vectors M_p and m_p near the equilibrium state, determined by expressions (2.65) and (2.66) presented above, we must again solve equations of motion of the form (2.52), written for each of the sublattices. For all the components, 12 equations are obtained. When they are linearized in the transverse (with respect to the equilibrium directions) components of the vectors M_p and m_p, there remains a system of eight homogenous linear equations.* The eighth-degree determinant corresponding to this system, the vanishing of which is the condition for the existence of nontrivial solutions, breaks up into two fourth-degree determinants. Two biquadratic equations are thereby obtained for the vibration eigenfrequencies sought. These equations can be written, to a very good approximation, in the following form (Turov and Kuleyev, 1965):

$$(\omega^2 - \omega_e^2)(\omega^2 - \omega_{n0}^2) - \omega^2\omega_T^2 = 0\,.$$

(2.67)

Here ω_e stands for two unperturbed (by the interaction with the nuclear spins) antiferromagnetic resonance frequencies (AFMR), which in the case under consideration are defined by the expressions

$$\omega_e = \omega_{e1} = |\gamma_e| H$$

(2.68)

or

$$\omega_e = \omega_{e2} = \omega_{Ea}\sqrt{1 - \left(\frac{H}{H_E}\right)^2}\,.$$

(2.69)

Also note that $\omega_{n0} = |\gamma_n H_n| = |\gamma_n A_0 M_0|$.

The "mixing" of the electron and the nuclear spin vibrations and the dynamic shift in their resonance frequencies, analogous to that considered above in the case of a ferromagnet, are caused by the presence in (2.67) of a term with the characteristic frequency

$$\omega_T = |\gamma_e|\sqrt{H_E|A_0 m_0|}\,.$$

(2.70)

---

* The remaining four equations are identically satisfied in the linear approximation by the already known static (longitudinal with respect to the equilibrium directions mentioned) components of the vectors M_p and m_p.

The magnitude of ω_T is determined by the geometrical mean value of the exchange field H_E and the mean HFI field $A_0 m_0(T)$ which is applied to the electrons by the nuclei. The latter can be written in the form

$$A_0 m_0 = A_0 \chi_n H_n = -\frac{C_0}{T}, \tag{2.71}$$

where

$$C_0 = \frac{I(I+1)}{3S} \frac{\hbar\omega_{n0}}{\kappa} \frac{\omega_{n0}}{|\gamma_e|}. \tag{2.72}$$

This expression for C_0 is multiplied by the relative concentration of the magnetic isotope, if the isotopic composition of the nuclei is only in part magnetic.

In favorable cases C_0 may attain a magnitude of the order of 10, as, for example, for the nuclei of the Mn^{++} ions, for which one can take $S = I = 5/2$ and $\omega_{n0} = 3 \cdot 10^9 \ sec^{-1}$. Consequently, in this case, the frequency ω_T at $1°K$ will be of the order of $10^{10}-10^{11} \ sec^{-1}$.

At high temperatures, when $\omega_T^2 \sim 1/T$ becomes very small, the last term in (2.67) can be neglected, and this equation gives two unperturbed frequencies (for each value of ω_e): $\omega = \omega_e$ and $\omega = \omega_{n0}$. In this case there is no dynamic coupling between the vibrations of the electron and the nuclear spins. On the contrary, at sufficiently low temperatures (when ω_T approaches in order of magnitude ω_{n0} or even ω_e) the coupling between the vibrations of the nuclear and electron spin-systems becomes very strong, and instead of ω_e and ω_{n0} in equation (2.67) we obtain two new frequencies of the common vibration spectrum: $\omega = \Omega_1$ and $\omega = \Omega_2$.

Before we proceed to discuss the solutions Ω_1 and Ω_2 of equation (2.67), we wish to point out that between these two solutions there exists the following simple relation:

$$\Omega_1 \Omega_2 = \omega_e \omega_{n0}. \tag{2.73}$$

The following "conservation law" is observed: the coupling between the vibrations of the electron and nuclear spins changes the eigenfrequencies of these vibrations in such a way that the product of these frequencies remains constant (in the same approximation in which equation (2.67) is valid).

Let us now consider approximate solutions of equations (2.67), the exact solutions of which have, in the general case, the form

$$\Omega_{1,2}^2 = \tfrac{1}{2}(\omega_e^2 + \omega_T^2 + \omega_{n0}^2) \pm \tfrac{1}{2}[(\omega_e^2 + \omega_T^2 + \omega_{n0}^2)^2 - 4\omega_e^2\omega_{n0}^2]^{1/2}. \tag{2.74}$$

The character of the approximation depends on the relation between the quantities ω_e, ω_T and ω_{n0}. For the case of the uniaxial crystal under consideration one of the antiferromagnetic resonance frequencies ω_{e2} defined by expression (2.69) is usually large in comparison with ω_T and ω_{n0}: $\omega_{e2} \gg \omega_T$ and ω_{n0}. Therefore, from (2.74)

we approximately find

$$\Omega_1 \simeq \omega_{e2}\left(1 + \frac{\omega_T^2}{2\omega_{e2}^2}\right) \equiv \Omega_{e2},$$

$$\Omega_2 \simeq \omega_{n0}\left(1 - \frac{\omega_T^2}{2\omega_{e2}^2}\right) \equiv \Omega_{n2}.$$

(2.75)

Here, the relative "temperature" shifts in the electron and nuclear frequencies, due to the dynamic coupling of the vibrations of these subsystems, are equal,

$$\frac{\delta\omega_{e2}}{\omega_{e2}} = -\frac{\delta\omega_n}{\omega_n} = \frac{1}{2}\frac{\omega_T^2}{\omega_{e2}^2} = \frac{1}{2}\frac{H_n}{H_a}\frac{m_0}{M_0},$$

(2.76)

i.e., they are of the same order of magnitude as in the case of a ferromagnet (compare with expression (2.60)). Consequently, the high-frequency branch of AFMR defined by (2.69) mixes weakly with the nuclear branches of the magnetic resonance.

At the same time, the AFMR frequency ω_{e1} (expression (2.68)) for the other, low-frequency vibration branch may, for not very high external magnetic fields $H(\sim 10^3 \text{ Oe})$ in the region of low temperatures, be comparable in magnitude with ω_T. If, however, it is assumed that

$$\frac{\omega_{n0}^2}{\omega_{e1}^2 + \omega_T^2} \ll 1,$$

then, expanding the radical in (2.74) in this parameter, we obtain in the first approximation

$$\Omega_1^2 \simeq \omega_{e1}^2 + \omega_T^2 \equiv \Omega_{e1}^2,$$

(2.77)

$$\Omega_2^2 \simeq \omega_{n0}^2 \frac{\omega_{e1}^2}{\omega_{e1}^2 + \omega_T^2} \equiv \Omega_{n1}^2.$$

(2.78)

Both these frequencies depend on the electron as well as on the nuclear characteristics. Nevertheless, the first of them (Ω_{e1}) can be called electron-like, and the second (Ω_{n1}) nuclear-like, because they are proportional to γ_e and γ_n respectively.

Let us point out that for the AFMR frequency ω_{e1} in the case of a small external field H, it is necessary to take into account the anisotropy in the basal plane. This means that in the initial expression (2.63) for the magnetic energy, it is, in addition, necessary to take into account terms of higher order in M_1 and M_2, invariant with respect to the symmetry transformations of the corresponding crystallographic class, which lead to the dependence of $\mathscr{H}$ on the direction in the basal plane of the antiferromagnetic vector $L = M_1 - M_2$ and the magnetization $M = M_1 + M_2$. Thereby instead of (2.68) we obtain (Turov, 1963)

$$\omega_e^2 = \omega_{e1}^2 = \gamma_e^2(H^2 + H_E H_\Delta),$$

(2.79)

where H_A is the effective magnetic anisotropy field in the basal plane, which depends on the orientation of H (and, consequently, of M) in this plane.* Taking these changes of ω_e into account, the expression for the frequencies of the coupled vibrations (2.77) and all subsequent formulas remain valid.

Comparing expressions (2.70), (2.77) and (2.79) it is not difficult to see that for the electron-like frequency (in the approximation under consideration) the HFI produces the same effect as an additional effective anisotropy field of magnitude $A_0 m_0$. Indeed expression (2.77) for Ω_{e1} is obtained from the unperturbed AFMR frequency (2.79) by the simple substitution

$$H_A \rightarrow H_A + |A_0 m_0| .$$

Equation (2.67) has a sufficiently simple form so as to allow the investigation of the dependence of the resonance field H on frequency ω and temperature T in the most general case for both branches of the spectrum. For this purpose it is necessary to solve (2.67) for $\omega_e \equiv \omega_e(H) \equiv \omega_H$:

$$\omega_H^2 = \omega^2 \left(1 - \frac{\omega_T^2}{\omega^2 - \omega_{n0}^2} \right). \tag{2.80}$$

Here we have introduced a new notation ω_H for the unshifted AFMR frequency ω_e in order to stress that the external field H enters relation (2.80) only through it. The dependence $\omega_H^2(\omega^2)$ is presented schematically in Figure 8. The resonant branch

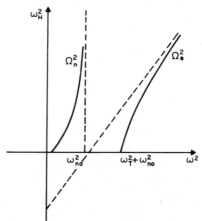

Figure 8

The frequencies of coupled vibrations of electron and nuclear spins in an antiferromagnet of EP type. The dependence of ω_H^2 on ω^2, corresponding to equation (2.80), is presented schematically.

---

* H_A can also be associated with magnetostrictive deformations; in this case it is isotropic in the first approximation (see Borovik-Romanov and Rudashevskii (1964), Turov and Shavrov (1965)). It must also be borne in mind that the simple expression (2.79) is only valid for $H^2 \gg H_E H_A$, when the magnetization $M \parallel H$.

in the frequency interval $\omega^2 < \omega_{n0}^2$ can be called nuclear-like (Ω_n), and the other branch, for frequencies $\omega^2 > \omega_T^2 + \omega_{n0}^2$—electron-like ($\Omega_e$). The frequency interval $\omega_{n0}^2 \leqq \omega^2 \leqq \omega_{n0}^2 + \omega_T^2$ is forbidden.

Since, according to (2.70) and (2.71) $\omega_T^2 = \text{const}/T$, expression (2.80) determines, for both branches of the spectrum simultaneously, the temperature dependence of the resonance field H (for a given oscillator frequency ω). It should be pointed out that here the temperature T is, in general, the nuclear spin-system temperature T_n. As has already been shown, it can, thanks to the NMR saturation effect, depend considerably on the intensity of the r.f. field h. Therefore, the position of both resonance frequencies will depend on the intensity of the r.f. field at the NMR frequency and on ω_T^2. However, according to (2.73) the product $\Omega_e \Omega_n$ remains unchanged.

The effect mentioned is used for the indirect observation of NMR by means of the phenomenon of double resonance.* If the magnitude of the resonance field H for the antiferromagnetic resonance (for a given frequency Ω_e) is measured as a function of the variable frequency of the second oscillator, then, for $\omega \sim \Omega_n$, a shift in H takes place, depending on the intensity of the r.f. field at the frequency ω. It is obvious that the frequency at which the shift in H is maximal is the NMR frequency Ω_n. Measuring the dependence of the NMR frequency thus obtained on the power of the r.f. field (i.e., on the temperature of the nuclear spin-system T_n) the unshifted NMR frequency ω_{n0} and, consequently the hyperfine field H_n can be found by extrapolation to $T_n \to \infty$.

The effects of the dynamic coupling between the electron and nuclear spin-system considered above are today well-known and have been investigated experimentally in a number of antiferromagnets, mainly manganese compounds. As it appears from the preceding, at low temperatures only the low frequency AFMR branches mix considerably with the nuclear spin vibrations. Antiferromagnets of the EP type, which were discussed above, have such a branch. A "pure" representative of such antiferromagnets is the compound $CsMnF_3$, a hexagonal crystal, antiferromagnetic below $53.5°K$ (Lee et al., 1963; Minkewicz and Nakamura, 1966b). The unshifted AFMR frequency ω_e for this compound must be taken in the form (2.79).

Some of the antiferromagnets of this type possess a weak ferromagnetism in the antiferromagnetic state, caused by a small deviation of moments M_1 and M_2 from strict parallel alignment, due to internal magnetic fields** (see Figure 7).

The best studied are the magnetic and resonant properties of manganese carbonate $MnCO_3$, which is a rhombohedral crystal, antiferromagnetic with a weak ferromagnetism below the Néel temperature of $32.5°K$ (Borovik-Romanov and Tulin, 1965; Shaltiel, 1966). The weak ferromagnetism in this compound is caused by the presence in the free energy (2.63) of an additional term (Dzyaloshinskii, 1957)

---

* Heeger et al. (1961), Lee et al. (1963), Fink and Shaltiel (1964).
** Dzyaloshinskii (1957), Borovik-Romanov (1962), Turov (1963), Moriya (1963).

$$D\left(M_{1x}M_{2y} - M_{2x}M_{1y}\right),\tag{2.81}$$

where D is a constant.

Repeating the calculation of the equilibrium magnetizations M_p and m_p with this term taken into account, we obtain instead of (2.66)

$$m_\perp \cong \chi_n\left(H - \tfrac{1}{2}A_0M_\perp\right),$$

$$M_\perp \cong \chi_\perp\left(H + H_D\right),\tag{2.82}$$

where $H_D = DM_0$ is the so-called Dzyaloshinskii field which is also the cause for the presence of a weak spontaneous magnetic moment at $H = 0$. All the expressions for the frequencies of the coupled vibrations (2.67), (2.73), etc., remain in force. It is only necessary to write now (Turov, 1963)

$$\omega_{e1} = |\gamma_e|\left[H_EH_A + H\left(H + H_D\right)\right]^{1/2},\tag{2.83}$$

$$\omega_{e2} = |\gamma_e|\left[H_EH_a + H_D\left(H + H_D\right)\right]^{1/2}\tag{2.84}$$

for the frequencies of the unshifted antiferromagnetic resonance ω_e.

Of course, the strong coupling between the vibrations of the electron and nuclear spins occurs not only in uniaxial antiferromagnets of the EP type, but also in general in antiferromagnets of arbitrary symmetry for which, for whatever reason, any of the antiferromagnetic resonance frequencies falls in a region of sufficiently low frequencies. This applies, for example, to cubic antiferromagnets for which the anisotropy is usually small, and both AFMR frequencies fall in the low frequency region. The best example of such an antiferromagnet is the compound $RbMnF_3$ (Heeger and Teaney, 1964), having a perovskite structure. Here one has to use for ω_e expressions of the form (2.79) and (2.69), where H_a and H_A are determined by the same cubic anisotropy. Therefore, in the given case, both branches of the antiferromagnetic resonance are of low frequency type and both of them "mix" with the nuclear spin vibrations at low temperatures. However, the following must be borne in mind.

As we shall see later, the conditions for the excitation of vibrations with the frequencies Ω_{n1} and Ω_{e1} (which are the solutions of equation (2.67) for $\omega_e = \omega_{e1}$) differ from the conditions for the excitation of vibrations with frequencies Ω_{n2} and Ω_{e2} (the solutions of the same equation (2.67) for $\omega = \omega_{e2}$). The first pair of frequencies is excited when the direction of the r.f. field is perpendicular with respect to the constant magnetizing field, $h \perp H$. The condition for the excitation of the second pair, however, is that the fields be parallel, $h \parallel H$. This difference moreover entails also the following. It can be shown that the NMR amplification factor at the frequency Ω_{n1} is large, $\eta \gg 1$, while at the frequency Ω_{n2} there is in fact no amplification, $\eta \sim 1$. Therefore, the observation of the second nuclear-like resonance is impeded

by the fact that, in the absence of strict parallel alignment of h and H, a signal may appear from the first resonance, which has considerably greater intensity.*

Another example of a weakly anisotropic antiferromagnet crystal in which a coupling was observed between the antiferromagnetic and nuclear resonances is another compound of the $XMnF_3$ type, namely $KMnF_3$ (Heeger et al., 1961; Witt and Portis, 1964a, b; Minkewicz and Nakamura, 1966a). In the paramagnetic phase the $KMnF_3$ crystal too has the perovskite structure; however, below the Néel point $(81.5°K)$, it experiences a tetragonal and subsequently, apparently, a rhombic distortion. A weak ferromagnetism thus appears.

For the rhombic symmetry, the energy causing the weak ferromagnetism has the form (Turov and Naish, 1960)

$$D_1 (M_{1x}M_{2y} - M_{2x}M_{1y}) - D_2 (M_{1x}M_{1y} - M_{2x}M_{2y}). \tag{2.85}$$

Repeating the calculation for the coupled vibrations of the electron and nuclear spins, we obtain, for this case again, an equation of the form (2.67) in which now

$$\omega_e = \omega_{e1} = |\gamma_e| \left[H_E H_{a1} + H (H + H_{D1} + 5H_{D2}) \right]^{1/2} \tag{2.86}$$

or

$$\omega_e = \omega_{e2} = |\gamma_e| \left[H_E H_{a2} + H (H_{D1} + H_{D2}) \right]^{1/2}. \tag{2.87}$$

These expressions are written for the case when the field H lies along the axis of the spontaneous weak ferromagnetic moment (X); here H_{a1} and H_{a2} constitute the two effective fields of rhombic magnetic anisotropy, while H_{D1} and H_{D2} are two Dzyaloshinskii fields, proportional to D_1 and D_2, respectively. For an arbitrary direction of H in the XY plane, the fields H_a and H_D will vary periodically (with a period of $180°$) as a function of the angle in this plane.**

Let us again note that equations (2.67), etc., for the frequencies of the coupled vibrations, as well as equation (2.80), which determines the dependence of the magnitude of the resonance field H on the frequency for both branches of the spectrum (see Figure 8), are suitable for all the types of crystals considered, when the antiferromagnetic resonance frequencies ω_e are assigned the appropriate values.

---

* In addition, in the majority of cases, the antiferromagnetic resonance at the second frequency ω_{e2} does not in fact depend on the magnitude of the external field (for $H \ll H_E$ in the case of (2.69) and for $H \ll (H_E H_a / H_D)$ in the case of (2.84)). This creates a definite difficulty also for the observation of the second electron-like resonance at the frequency Ω_{e2}.

** We have previously assumed that the field H is sufficiently large so that the antiferromagnetic vector L remains perpendicular to H while the latter is rotated. Let us also note that under the square root in expression (2.86) and (2.87) there should be other terms, not depending on H and quadratic in H_{D1} and H_{D2}. We assume that they are included in the first term (with H_{a1} or H_{a2}). In the cited works (Heeger et al., 1961; Witt and Portis, 1964a; Minkewicz and Nakamura, 1966a) the frequencies of antiferromagnetic resonance (2.86) and (2.87) for $KMnF_3$ are written on the assumption that $|D_2| \gg |D_1|$ and it is in fact assumed that $H_{D1} = 0$.

It is interesting that in the case of strong coupling, when ω_T is comparable with ω_{e1}, the electron-like resonance frequency Ω_{e1} (or for a given frequency, the resonance field) becomes strongly temperature dependent, while the nuclear-like frequency Ω_{n1} becomes strongly dependent on the magnitude of the field H (the latter is seen again in Figure 8). As an illustration of the first assertion, the temperature dependence of the resonance field (for the electron-like branch) is shown in Figure 9a, for the hexagonal antiferromagnet $CsMnF_3$, as obtained by Lee et al. (1963) for two directions in the basal plane. The dependence of the nuclear-like resonance frequency on the external field, appearing at low temperatures, is clearly seen in the curves of Figure 9b. This figure refers to the rhombohedral weakly ferromagnetic compound $MnCO_3$ and is taken from the work of Borovik-Romanov and Tulin (1965).

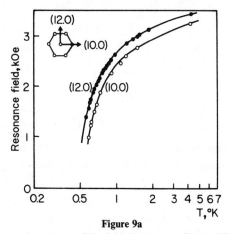

Figure 9a

The temperature dependence of the field H corresponding to AFMR in $CsMnF_3$.

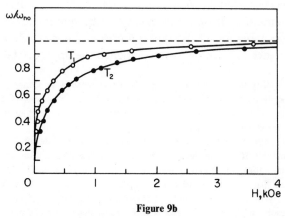

Figure 9b

The field dependence of the NMR frequency of Mn^{55} in $MnCO_3$ at temperatures $T_1 = 4.2°$ and $T_2 = 1.8°K$.

The investigation of nonlinear phenomena carried out on the manganese compounds enumerated above (mainly by the use of electron-nuclear double resonance) has revealed a number of unexpected effects. It turned out that saturation is observed at a given temperature even at frequencies ω separated from the shifted resonance frequency Ω_{n1} by an amount substantially exceeding the line width. For this it is only necessary that the frequency ω should fall in the interval $\Omega_{n1} < \omega < \omega_{n0}$ between the shifted and the unshifted frequencies and that the power of the r.f. field should attain some threshold value. Witt and Portis (1946b) consider that volume inhomogeneities in the crystal serve as nucleation centers for partial saturation, which can propagate over the whole specimen when the threshold power is reached.

Another interesting nonlinear effect, observed by double resonance, is the hysteresis in the electron-branch resonance field as a function of the r.f. power, at the frequency of the nuclear branch (Heeger et al., 1961; Witt and Portis, 1964b).

Up to now we have considered the case of the transverse magnetization of an antiferromagnet when the magnetizing field H was perpendicular to the antiferromagnetic vector L. The latter is typical of an antiferromagnet of the EP type for which a comparatively weak field in the "easy plane" induces the state with $L \perp H$.

In the case of a uniaxial antiferromagnet with the antiferromagnetic axis lying initially along the main axis of the crystal (and also for rhombohedral crystals), the magnitude of the threshold field $H_n \sim \sqrt{H_E H_a}$ for which "spin flop" of the antiferromagnetic vector to the state with $L \perp H$ occurs (assuming that initially $L \parallel H \parallel Z$) usually reaches tens and hundreds of kiloersteds. Typical antiferromagnets of this type are MnF_2, CoF_2 and Cr_2O_3 (see, for example, Borovik-Romanov, 1962).

In the uniaxial case with longitudinal magnetization ($H \parallel Z$), the NMR frequencies of these antiferromagnets (with the dynamic shift taken into account) have the approximate form (De Gennes et al., 1963)

$$\Omega_n^{(1,2)} = \gamma_n |H_n| \left(1 - \frac{\omega_T^2}{\omega_{e1}\omega_{e2}} \right)^{1/2} \pm \gamma_n H \left(1 + \frac{\gamma_e^2 A_0^2 M_0 |m_0|}{\omega_{e1}\omega_{e2}} \right), \qquad (2.88)$$

where now the unshifted NMR frequencies take the values (see expression (2.35))

$$\omega_{e(1,2)} = |\gamma_e| \left[\sqrt{H_E H_a} \pm H \right]. \qquad (2.89)$$

In the given case, vibrations of all these frequencies (NMR and AFMR) are excited by an r.f. field perpendicular to the constant field.

To orders of magnitude, the relative dynamic frequency shift is again reduced to an expression of the form (2.76) (provided H is not very close to the value of the threshold field $\sqrt{H_E H_a}$), so that this shift is as small here as for ferromagnets.

We see from (2.88) that the degeneracy of the two NMR frequencies is lifted only by the external field. Note that one of these frequencies is excited by a right polarized circular component of the r.f. field, and the other, by a left polarized component.

CHAPTER III

Radiofrequency Susceptibility. Amplification Coefficient

In order to explain in detail the mechanism of NMR amplification in ferromagnets and antiferromagnets, and also to ascertain the conditions ("selection rules") for the excitation of different types of vibrations, it is necessary to examine the behavior of a system of electron and nuclear spins in a r.f. field h.

1. Ferromagnets

In the case of ferromagnets, the spectrum of which was considered in Chapter II, Sec. 3, results of interest to us can be obtained in the simplest manner if we assume that a rotating magnetic field $h \perp H$ acts on the system,

$$h_x = h \cos \omega t, \quad h_y = h \sin \omega t, \quad h_z = 0. \tag{3.1}$$

The equations of motion (2.52) in the linear approximation are thus again reduced to the form (2.53) with the only difference that in the right-hand side of (2.53) and (2.53a) the following terms are added:

$$\mp i\gamma_e M h \exp(\pm i\omega t) \quad \text{and} \quad i\gamma_n m h \exp(\pm i\omega t).$$

The solution of such a system of equations will have the form

$$M^\pm = M_h \exp(\pm i\omega t), \qquad m^\pm = m_h \exp(\pm i\omega t), \tag{3.2}$$

where

$$M_h = \chi_M h, \quad m_h = \chi_m h. \tag{3.3}$$

χ_M and χ_m, constituting one of the circular components of the r.f. susceptibility tensor of the electron and nuclear subsystem respectively, are defined by the following expressions:

$$\chi_M = \frac{|\gamma_e|\, M\, [\Omega_n - \omega + (\eta + 1)\,\gamma_n A_0 m]}{(\omega - \Omega_e)(\omega - \Omega_n)}, \tag{3.4}$$

$$\chi_m = \frac{\gamma_n m(\omega - \Omega_e) + |\gamma_e|\,\gamma_n A_0 M m}{(\omega - \Omega_e)(\omega - \Omega_n)}. \tag{3.5}$$

Here Ω_e and Ω_n are the FMR and NMR frequencies, taking into account the shifts given by expressions (2.58) and (2.59); $\eta = H_n/(H + H_a)$ is the NMR amplification coefficient, introduced in Chapter I (expression (1.9)).

Let us note that for the observation of NMR it is neccessary that the sign of ω (which determines the sense of rotation of the r.f. field) should coincide with the sign of the product $\gamma_n A_0$ (for $H < |H_n|$). In spite of the fact that the magnetizations M and m usually have opposite signs, the NMR and FMR must be observed with r.f. fields rotating in the same sense since, in the majority of cases, the gyromagnetic ratios of the nucleus and of the electron also have opposite signs.

Expressions (3.4) and (3.5) near the NMR frequency $\omega \sim \Omega_n \ll \Omega_e$ take the form

$$\chi_M \cong \chi_e - \chi_n \frac{\omega_n}{\omega - \Omega_n}\,\eta\,(\eta + 1), \tag{3.6}$$

$$\chi_m \cong -\chi_n \frac{\omega_n}{\omega - \Omega_n}(\eta + 1), \tag{3.7}$$

where $\chi_e = M/(H + H_A)$ is the static electron susceptibility coinciding with the rotation susceptibility (see p.7), χ_n is the static nuclear susceptibility, ω_n is the NMR frequency without taking into account the dynamic shift (see expression (2.57)).

Note that, according to (3.6) and (3.7), because of the condition $|\eta| \gg 1$ it is the electronic part of the susceptibility that makes the main contribution to the total dispersion curve of the whole system

$$\chi(\omega) = \chi_M + \chi_m = \text{const} + \chi_n \frac{\omega_n}{\Omega_n - \omega}(1 + \eta)^2 \tag{3.8}$$

even near the NMR frequency. This conclusion can be carried over also to the absorption curve.

As a matter of fact, the absorption can formally be taken into account by introducing into the denominator of expression (3.8) for the susceptibility χ (or in the corresponding expressions for χ_M and χ_m) the complex resonance frequency

$$\Omega_n \to \Omega_n + i\Gamma_n, \tag{3.9}$$

where Γ_n is the half-width of the resonance line. We thereby obtain

$$\chi = \chi' - i\chi'',$$

where

$$\chi' = \text{const} + \chi'_n (1 + \eta)^2, \tag{3.10}$$

$$\chi'' = \chi''_n (1 + \eta)^2, \tag{3.11}$$

$$\chi'_n = \chi_n \frac{\omega_n (\Omega_n - \omega)}{(\Omega_n - \omega)^2 + \Gamma_n^2}, \tag{3.12}$$

$$\chi''_n = \chi_n \frac{\omega_n \Gamma_n}{(\Omega_n - \omega)^2 + \Gamma_n^2}. \tag{3.13}$$

In expression (3.11) for χ'', as well as in (3.10) for χ', the main term ($\sim \eta^2$) is of electronic origin.

In nuclear resonance the energy is not absorbed directly by the nuclear spins in the system but mainly via the electrons as a result of the resonant modulation of their motion by the interaction with the nuclei. From the quantum mechanical point of view this means that the main contribution to the probability of nuclear dipole transitions comes from the electron term of the interaction energy with the r.f. field (i.e., $-\gamma_e \hbar (S_x h_x + S_y h_y)$). This is made possible thanks to the "mixing" of various spin states of the whole system (with the same total magnetic quantum number), caused by the hyperfine interaction between the nuclei and the electrons (see, for example, Abragam, 1963).*

The results for χ obtained above describe the amplification of NMR in a ferromagnet as an amplification by a factor $(1 + \eta)^2$ of the r.f. susceptibility at the NMR frequency. At the same time, since the absorbed power is

$$\mathcal{P} \sim \chi'' h^2 \equiv \chi''_{\eta=0} [(\eta + 1) h]^2, \tag{3.14}$$

the amplification can be related to the r.f. field ($h \to (1 + \eta) h \approx \eta h$), in accordance with the qualitative treatment of this problem in Chapter I, while the r.f. susceptibility is considered the same as in a diamagnetic crystal.

Let us point out, however, that sometimes the resonant absorption near the NMR frequency has the form of the dispersion curve χ'_n (3.12) and not of the absorption curve χ''_n (3.13) (Portis and Gossard, 1960; Review 1965b). It is not difficult to understand this point if in the initial system of equations (2.52), in addition to the attenuation of the nuclear spins, the attenuation of the electron spins is also taken into account. Formally the matter is reduced to the following substitution in expressions

---

* The assumption of considerable amplification of the nuclear resonance line intensity at the nuclei of paramagnetic atoms (as compared with diamagnetic atoms) due to hyperfine coupling, was apparently first expressed in the work of Valiev (1957).

(3.4) and (3.5):

$$\Omega_e \to \Omega_e + i\Gamma_e.$$

Here Γ_e is a parameter characterizing the relaxation of the electron magnetization. This substitution together with (3.9) leads, after the elimination of the imaginary part in the full susceptibility $\chi = \chi_M + \chi_m$, instead of (3.11) to an expression of the form (near the frequency $\omega \sim \Omega_n \ll \omega_e$ and for $|\eta| \gg 1$)

$$\chi'' \cong \chi_e \frac{\Gamma_e}{\omega_e} + \eta^2 \left(\chi_n'' + \frac{\Gamma_e}{\omega_e} \chi_n' \right). \tag{3.15}$$

As we see from (3.15), NMR signals from both the absorption and the dispersion curves, amplified by a factor of η^2 and $\eta^2 (\Gamma_e/\omega_e)$, respectively, are generally superimposed against the nonresonant background of energy losses in the electron spin-system.*

The resulting signals may have a dispersive character even for $\Gamma_e \ll \omega_e$ in the case when the NMR saturation condition (2.62) is fulfilled. In saturation, as is known (see, for example, Abragam, 1963) χ_n'' decreases with the increase of the r.f. field amplitude considerably faster than χ_n'.

The effects mentioned will be considerable particularly in the case of NMR on nuclei in domain walls, for which the amplification factor η and the relative attenuation Γ_e/ω_e are usually larger than for the domains themselves.

In the foregoing, the phenomenological treatment of the susceptibility

$$\chi = \chi_M + \chi_m$$

was carried out only with regard to a circularly polarized r.f. field (3.1). It is not difficult to show that if the system has axial symmetry, which is the case here (the Z axis, parallel to the constant field H, is at the same time the symmetry axis of the crystal), it is possible to express all nine components of the r.f. susceptibility tensor $\chi_{\alpha\beta}(\omega)$ through this circular susceptibility $\chi(\omega)$. We have in this case

$$\chi_{xx} = \chi_{yy} = \tfrac{1}{4} \left[\chi(\omega) + \chi^*(-\omega) \right],$$
$$\chi_{yx} = -\chi_{xy} = \frac{1}{4i} \left[\chi(\omega) - \chi^*(-\omega) \right], \tag{3.16}$$
$$\chi_{zx} = \chi_{zy} = \chi_{yz} = \chi_{zz} = \chi_{xz} = 0.$$

---

* Isolating the real part of the full susceptibility $\chi_M + \chi_m$ it is not difficult to show that its main terms can be written in the form

$$\chi' \simeq \chi_e + \eta^2 \left(\chi_n' - \frac{\Gamma_e}{\omega_e} \chi_n'' \right).$$

Here, conversely, the absorption curve admixes with the NMR dispersion curve.

It should be borne in mind, however, that the components in the second line of (3.16) do not have a resonant character; near the NMR frequency they can be neglected. Thus, the behavior of the electron-nuclear spin-system in a ferromagnet near the NMR frequency in a linearly polarized r.f. field is also determined in the main only by the circular susceptibility $\chi(\omega)$.

Further analysis of the problem of the NMR amplification coefficient and the superposition of dispersion and absorption signals will be carried out in Chapter V on the basis of the microscopic theory of NMR susceptibility.

2. Antiferromagnets

An analogous, though more complex, susceptibility calculation can be carried out also for antiferromagnetism. We shall present here results of such a calculation, again for the case when the antiferromagnetic axis lies in the XY plane with small anisotropy (Turov and Kuleev, 1965).

Expressions (2.74) define in this case two pairs of resonance frequencies. It turns out, accordingly, as has already been mentioned before, that the first pair of frequencies $\Omega_1 \equiv \Omega_{e1}$ and $\Omega_2 \equiv \Omega_{n1}$ are excited when the r.f. field h is perpendicular to the constant magnetizing field H (i.e., $h \perp H \| X$; see Figure 7), while the second pair $\Omega_1 \equiv \Omega_{e2}$ and $\Omega_2 \equiv \Omega_{n2}$ are excited when h and H are parallel.

The main terms in the susceptibilities with respect to the r.f. field $h \perp H$, for the cases $h \| Y$ and $h \| Z$ respectively, have the following form:

$$\chi_{yy} \cong \chi_{yy}^M = \chi_{e\perp} \frac{(\omega_{e1}^2 - \omega_A^2)(\omega_{n0}^2 - \omega^2)}{(\Omega_{e1}^2 - \omega^2)(\Omega_{n1}^2 - \omega^2)}, \tag{3.17}$$

$$\chi_{zz} \cong \chi_{zz}^M = \chi_{ez} \frac{\Omega_{e1}^2 \Omega_{n1}^2 - \omega^2(\omega_{e1}^2 + \omega_T^2)}{(\Omega_{e1}^2 - \omega^2)(\Omega_{n1}^2 - \omega^2)}. \tag{3.18}$$

Here $\chi_{e\perp}$ and χ_{ez} are the static susceptibilities for the directions $H \perp Z$ and $H \| Z$ respectively, obtainable as the ratio of the static magnetization M in the corresponding direction to the magnetizing field H. In particular, in the presence of weak ferromagnetism

$$\chi_{e\perp} = \frac{M_\perp}{H} = \chi_\perp (H + H_D)/H \tag{3.19}$$

and $\chi_{ez} = M_z/H \cong \chi_{\perp z}$ where $\chi_\perp$ and $\chi_{\perp z}$ are the transverse magnetic susceptibilities of the antiferromagnet in the corresponding directions. Finally, ω_A is the maximum value of the AFMR frequency when extrapolated to $H = 0$; for example, for $MnCO_3$, $RbMnF_3$ and $CsMnF_3$, $\omega_A^2 = \gamma_e^2 H_E H_A$, where H_A is the anisotropy field in the XY plane.

Near the NMR frequency expressions (3.17) and (3.18) can be rewritten as

$$\chi_{yy} = \chi_n \eta_y^2 \frac{\Omega_{n1}^2}{\Omega_{n1}^2 - \omega^2}, \qquad (3.20)$$

and

$$\chi_{zz} = \chi_n \eta_z^2 \frac{\Omega_{n1}^2}{\Omega_{n1}^2 - \omega^2}, \qquad (3.21)$$

where χ_n is the static susceptibility of the nuclear spin-system, which we have encountered before (see expression (2.66)), while η_y and η_z are the NMR amplification coefficients defined by the expressions

$$\eta_y^2 = \left(\frac{H_n}{H}\right)^2 \frac{(\omega_{e1}^2 - \omega_\Delta^2)^2}{\Omega_{e1}^2 \omega_{e1}^2}, \qquad (3.22)$$

$$\eta_z^2 = \left(\frac{\gamma_e H_e}{\Omega_{e1}}\right)^2 \left(\frac{\omega_{n0}}{\Omega_{e1}}\right)^2. \qquad (3.23)$$

As ought to be expected, $\eta_y^2 \gg \eta_z^2$ since, as in the case of ferromagnets, $\eta_y \sim H_n/H$ (for not very low temperatures, when $\omega_T \lesssim \omega_{e1}$). This is understood from simple physical considerations: the effective r.f. field which acts on the nuclear spin via HFI is, in the given case (small anisotropy in the basal plane), defined as

$$h_y^* \sim A_0 M_0 \theta \sim A_0 M_0 \frac{h_y}{H} = \frac{H_n}{H} h_y. \qquad (3.24)$$

Here θ is the angle by which the antiferromagnetic vector L turns under the action of h_y.

Note that for the limiting case of low temperature ($\omega_T^2 \gg \omega_{e1}^2$), if also $\omega_{e1}^2 \gg \omega_\Delta^2$, the square of the amplification coefficient is simply equal to the ratio of the electron and nuclear static susceptibilities: $\eta_y^2 \cong \chi_{e\perp}/\chi_n$.

The amplification of the r.f. field, directed perpendicular to the plane in which the magnetic moments of the sublattices and the constant field H lie, depends according to (3.23) on the ratio of the NMR and AFMR frequencies.

As has already been mentioned before (see Chapter II, Sec. 3) there is usually no amplification for the second NMR branch Ω_{n2} excited when $h\|H\|X$. This is seen from the chain of relations of the form (3.24) in which, in the given case, the angle $\theta \sim h_x/H_E$ and consequently $\eta_x \sim H_n/H_E$.*

It is interesting that expression (3.17) leads not only to NMR amplification but also to the conclusion that there is some weakening of the AFMR intensity. As a

---

* All this refers also to both NMR frequencies for an antiferromagnet of the "easy axis" type (see below).

matter of fact, near the frequency $\omega \sim \Omega_{e1}$ we have

$$\chi_{yy} \cong \chi_{e\perp} \frac{\omega_{e1}^2}{\Omega_{e1}^2 - \omega^2} = \zeta^2 \chi_{e\perp} \frac{\Omega_{e1}^2}{\Omega_{e1}^2 - \omega^2} \tag{3.25}$$

(for the sake of simplicity we have again assumed $\omega_{e1}^2 \gg \omega_{n0}^2, \omega_\Delta^2$) where the quantity

$$\zeta = \frac{\omega_{e1}}{\Omega_{e1}} \cong \left(1 - \frac{\omega_T^2}{\Omega_{e1}^2}\right)^{1/2} \tag{3.26}$$

can be called the AFMR attenuation coefficient. It is easy to see that it can, for the given frequency $\omega = \Omega_{e1}$, lead to a reduction in the AFMR intensity at low temperatures. Indeed, by substituting $\Omega_{e1} \rightarrow \Omega_{e1} + i\Gamma_e$ in the denominator of expression (3.25), we obtain for the imaginary part of the susceptibility at resonance ($\omega = \Omega_{e1}$)

$$\chi_{yy}'' = \zeta^2 \chi_{e\perp} \left(\frac{\Omega_{e1}}{2\Gamma_e}\right).$$

Thus, in the known relation for the relative height of the resonance peak

$$\frac{\chi_{max}''}{\chi_0} = \frac{\omega_{res}}{\Delta\omega}$$

(where χ_0 is the static susceptibility and $\Delta\omega = 2\Gamma$ is the width of the resonance line) there appears, in the given case, a factor $\zeta^2 \leqslant 1$ on the right-hand side. It is possible that the decrease in the AFMR line intensity with the lowering of temperature, observed down to very low temperatures ($T \leqslant 1°K$) in $CsMnF_3$ (Lee et al., 1963) and $MnCO_3$ (Fink and Shaltiel, 1964), is associated with this effect.

In conclusion, note that analogous considerations for the two NMR frequencies (2.88) in the case of antiferromagnets of the EA type lead to an expression of the form (3.20) for the r.f. susceptibility, in which η_y^2 is replaced by a factor $(1 + H_n/H_E)$. In this case there will be amplification only for $|H_n| > H_E$, just as in antiferromagnets of the EP type for the resonance with the frequency Ω_{n2}.

CHAPTER IV

Indirect Interaction between Nuclear Spins. Nuclear Spin Waves

1. The Suhl–Nakamura Interaction

The interaction between nuclear spins and electrons, besides considerably affecting the NMR frequency, leads to a number of specific features in the behavior of the nuclei in ferromagnets and antiferromagnets. One of the most significant features is the peculiar indirect interaction of the nuclear spins via the electrons.

In metallic ferromagnets, as in ordinary (nonferromagnetic) metals, the nuclei interact, of course, indirectly through the intermediacy of the conduction electrons (the mechanism of Ruderman and Kittel, 1954). A characteristic feature of this interaction between the nuclear spins I_1 and I_2 is its pseudoexchange form

$$\mathcal{H}_{RK} = U_{RK}(R_{12})I_1 I_2 .$$

The interaction parameter $U_{RK}(R_{12})$ is of the order of A^2/E_F, where E_F is the Fermi energy.

The interaction $\mathcal{H}_{RK}$ does not contribute to the second moment of the NMR line (for the case of identical nuclei considered here), but it does affect the fourth moment, leading to the so-called pseudoexchange narrowing of the line.

The indirect interaction between nuclear spins via the conduction electrons is isotropic because in its calculation the electron spin polarization has been neglected. This approximation is possibly valid also for a ferromagnet, since the conduction electron spin polarization is small. In ferromagnets and antiferromagnets, however, the nuclei can interact via the inner shell electrons of the magnetic atoms, the strong magnetic polarization of which is in fact responsible for ferromagnetism or anti-

61

ferromagnetism. As has already been mentioned before (Chapter I, Sec. 2), this interaction can be described at low temperatures as an exchange of spin waves between the nuclei of different atoms (the emission of an electron spin wave by one nucleus and its absorption by another). This was first shown by Suhl (1958) for a ferromagnet and by Nakamura (1958) for an antiferromagnet.

The hyperfine interaction for the whole crystal (1.2) can be represented in the form

$$\mathscr{H}_{HFI} = A \sum_j I_j^z S_j^z + \frac{A}{2} \sum_j (I_j^+ S_j^- + I_j^- S_j^+). \qquad (4.1)$$

We shall first consider the case of a ferromagnet and introduce the creation and destruction operators for spin deflection $a^+(R_j)$ and $a(R_j)$ according to equation (2.12).

If the dipole term $\mathscr{H}'_{dip}$ in the Hamiltonian of the electron spin-system (2.11) is neglected, the remaining expression, which is quadratic in the operators a^+ and a, can be diagonalized (see (2.13)) in the case of a ferromagnet through a simple Fourier transformation

$$a(R_j) = \frac{1}{\sqrt{N}} \sum_k a_k e^{ikR_j}, \quad a^+(R_j) = \frac{1}{\sqrt{N}} \sum_k a_k^+ e^{-ikR_j}, \qquad (4.2)$$

where N is the number of electron spins. We can, therefore, substitute in the Hamiltonian $\mathscr{H}_{HFI}$ (4.1) the spin wave creation and destruction operators a_k^+ and a_k for the operators S_j^z using the relations (2.12) and the transformation (4.2). Consequently, the full Hamiltonian of the nuclear spin-system and of the electron spin waves can be written as a sum of two terms,

$$\mathscr{H} = \mathscr{H}_0 + \mathscr{H}', \qquad (4.3)$$

where the unperturbed Hamiltonian,

$$\mathscr{H}_0 = -(AS + \gamma_n \hbar H) \sum_j I_j^z + \sum_k \hbar \omega_k a_k^+ a_k \qquad (4.4)$$

is the sum of the nuclear spin energy in the effective field $H + (AS/\gamma_n \hbar)$ and of the electron spin wave energy ($\varepsilon_k = \hbar \omega_k$). The second term,

$$\mathscr{H}' = \frac{A}{N} \sum_{jkk'} I_j^z a_k^+ a_{k'} e^{i(k-k')R_j} + \frac{1}{2} A \left(\frac{2S}{N}\right)^{1/2} \sum_{jk} (I_j^- a_k^+ e^{-ikR_j} + I_j^+ a_k e^{ikR_j}) \qquad (4.5)$$

can be considered as a small perturbation determining the correction to the eigenvalues of $\mathscr{H}_0$.

Let us calculate the correction $\Delta E_{0\alpha}$ to the energy corresponding to $\Psi_{0\alpha} = \psi_0 \phi_\alpha$, where ψ_0 is the ground state wave function of the electron subsystem (i.e., the state with no spin waves) and ϕ_α is the wave function of an arbitrary state α of the nuclear subsystem. The excited state wave function of the electron subsystem is denoted by

ψ_n; in particular for the state with one spin wave of wave vector k we have $n = 1_k$. Since the diagonal matrix elements

$$(0 \,|\, a_k^+ a_{k'} \,|\, 0) = (0 \,|\, a_k^+ \,|\, 0) = (0 \,|\, a_k \,|\, 0) = 0 \tag{4.6}$$

vanish, there is no first order correction to the energy, i.e., $\Delta E_{0\alpha}^{(1)} = 0$. In the second order perturbation approximation, the correction due to the first term in (4.5) vanishes too (since $a_k \psi_0 = 0$), while that due to the remaining terms can be written in the form,

$$\Delta E_{0\alpha}^{(2)} = - \sum_{\alpha', k} \frac{(0\alpha \,|\, \mathcal{H}' \,|\, \alpha' \, 1_k)(\alpha' \, 1_k \,|\, \mathcal{H}' \,|\, 0\alpha)}{\hbar\omega_k + E_{\alpha'} - E_\alpha} \simeq$$

$$\simeq \frac{1}{2} \frac{SA^2}{N} \sum_{\alpha', j, j', k} \frac{(\alpha \,|\, I_j^+ \,|\, \alpha')(\alpha' \,|\, I_{j'}^- \,|\, \alpha) \, e^{ik(R_j - R_{j'})}}{\hbar\omega_k}. \tag{4.7}$$

(In the last link of (4.7) we have taken into account that generally $\hbar\omega_k \gg |E_{\alpha'} - E_\alpha|$). The last expression can be represented as the diagonal matrix element $\Delta E_{0\alpha}^{(2)} = (\alpha \,|\, \mathcal{H}_{\text{eff}} \,|\, \alpha)$ of an effective Hamiltonian,

$$\mathcal{H}_{\text{eff}} = \tfrac{1}{2} \sum_{j, j'} U_{jj'} I_j^+ I_{j'}^-, \tag{4.8}$$

where

$$U_{jj'} = - \frac{SA^2}{N} \sum_k \frac{e^{ik(R_j - R_{j'})}}{\hbar\omega_k}. \tag{4.9}$$

2. Nuclear Spin Waves in Ferromagnets

Since $\Delta E_{0\alpha}^{(2)}$ is its diagonal matrix element, the operator $\mathcal{H}_{\text{eff}}$ represents to the first approximation some effective interaction in the nuclear spin subsystem.

The full Hamiltonian of the latter, including the corresponding term of $\mathcal{H}_0$ (4.4), now becomes

$$\mathcal{H}_{\text{I}} = -\left[AS\left(1 + \frac{1}{2N} \sum_k \frac{A}{\hbar\omega_k}\right) + \gamma_n\hbar H \right] \sum_j I_j^z -$$

$$- \frac{1}{2} \frac{AS}{N} \left(\sum_k \frac{A}{\hbar\omega_k}\right) \sum_j [(I_j^x)^2 + (I_j^y)^2] + \frac{1}{2} \sum_{j \neq j'} U_{jj'} I_j^+ I_{j'}^-. \tag{4.10}$$

Here, the terms corresponding to $j = j'$ in (4.8) have been written separately (taking into account that $I^+ I^- = (I^x)^2 + (I^y)^2 + (I^z)^2$). The first of these terms is proportional

to $\sum_j I_j^z$ and it leads simply to a renormalization of the constant A in the first term of (4.10). The second term, however, is proportional to $\sum_j [(I_j^z)^2 - I(I + 1)]$ and thus, apart from a constant, it has the form of a quadrupole Hamiltonian for an electric field gradient tensor of cylindrical symmetry. The role of "quasi-quadrupole" terms will be dealt with in Chapter VI in the course of the discussion on quadrupole terms. In any case, for $I = 1/2$ it becomes a constant and has no significance.

Taking this into consideration, the effective Hamiltonian (4.10) becomes

$$\mathscr{H}_I = - (AS + \gamma_n \hbar H) \sum_j I_j^z + \tfrac{1}{2} \sum_{j \neq j'} U_{jj'} I_j^+ I_{j'}^- . \tag{4.11}$$

Now, if the quantum-mechanical equation of motion for I_j^+ (or I_j^-) is written as

$$\frac{dI_j^+}{dt} = \frac{i}{\hbar} [\mathscr{H}_I, I_j^+] , \tag{4.12}$$

the commutator $[\mathscr{H}_I, I_j]$ is calculated, and equation (4.12) is linearized (by replacing the operator I_j^z on the right-hand side with its thermodynamic mean value $\langle I_j^z \rangle \equiv I_z$), the following expression is obtained for the eigenfrequency spectrum of the nuclear spin-system (De Gennes et al., 1963):

$$\Omega_{nk} = - \left(\frac{AS}{\hbar}\right) \left(1 - \frac{AI_z}{\hbar \omega_k}\right) - \gamma_n H . \tag{4.13}$$

In the case of uniform oscillations (i.e., for $k = 0$), expression (4.13) coincides with the phenomenologically calculated NMR frequency shift (2.59) (if the relations $\omega_k (k = 0) \equiv \omega_e = |\gamma_e| (H + H_a)$, $m = \gamma_n \hbar I_z N$ and $A = A_0 \gamma_e \gamma_n \hbar^2 N$ are taken into account). Here the frequency shift proportional to I_z appeared as a result of the indirect interaction of nuclei via spin waves (see second term in (4.11)).

Expression (4.13), however, contains an additional result; it describes a quasi-continuous frequency spectrum (in an interval of width $\delta \omega_n = A^2 S |I_z|/\hbar^2 \omega_e$) which depends on the wave vector k (Figure 10). These waves, which are associated with the correlation of nuclear spin movements caused by the Suhl–Nakamura interaction, can be called *nuclear spin waves* (NSW).

The dispersion of NSW becomes ignorable (and their frequency $\Omega_{nk} \to \omega_n$) for such small values of k when the energy of the electron spin waves $\hbar \omega_k$ given by (2.21) can be expanded as

$$\omega_k = \omega_e + \omega_E (ak)^2 \tag{4.14}$$

(where a is the interatomic distance and ω_E is the exchange interaction parameter).

The critical value of the wave vector k, which divides between the regions of strong and weak shift in the NSW frequencies, can be found from the condition $\omega_k \sim 2\omega_e$ (i.e., as the value $k = k_0$ which corresponds to the middle of the NSW

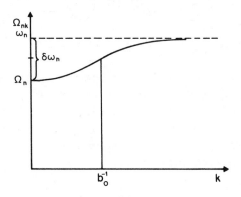

Figure 10

The spectrum of nuclear spin waves in a ferromagnet.
$k_0 = b_0^{-1}$ corresponds to the middle of the dispersion band (between the frequencies Ω_n and ω_n).

frequency band). This takes place for

$$k_0^{-1} = b_0 \sim a\,(\omega_E/\omega_e)^{1/2} \gg a. \tag{4.15}$$

Assuming, for example, $\omega_E/\omega_e \sim 1000$ we obtain $b_0 \sim 30\,a$.

It is easy to see that b_0 constitutes the effective radius of the Suhl–Nakamura interaction. Indeed, replacing in (4.9) the summation over k by integration and taking (4.14) into account, it is possible to express the quantity $U_{jj'} \equiv U(R_{jj'})$ which characterizes this interaction in the approximate form

$$U(R_{jj'}) = -\frac{A^2 S}{4\pi\hbar\omega_E}\,\frac{a}{R_{jj'}}\,\exp\left\{-\frac{R_{jj'}}{b_0}\right\}. \tag{4.16}$$

Thus, the interaction between nuclear spins via spin waves is appreciable only over distances $R \lesssim b_0$ and, at the same time, the NSW frequency is shifted relative to ω_n only for waves with wavelengths $\lambda \lesssim b_0$.

For other details of properties of NSW, these new interesting elementary spin excitations (propagation velocity and mean free path, contribution to heat capacity), we refer the reader to the original works of De Gennes et al. (1963).

It is necessary, however, to clarify here whether there is any sense in speaking about the existence of such collective vibrations of the nuclear spins at moderately low temperatures, for which $|I_z| \ll I$ (helium temperatures). The point is that for such temperatures the linearization procedure employed in the solution of equation (4.12) has to be justified. For this purpose, let us write down the spectral function (De Gennes et al., 1963; Pincus, 1963)

$$\mathscr{P}_k(\omega) = \int dt\,\langle I_k^-(0) I_k^+(t)\rangle\,e^{i\omega t}, \tag{4.17}$$

where

$$I_k^+(t) = \left(\exp\left(\frac{i}{\hbar} \mathscr{H}_I t \right) \right) \left(\sum_j I_j^+ \, e^{ikR_j} \right) \left(\exp\left(-\frac{i}{\hbar} \mathscr{H}_I t \right) \right)$$

and the thermodynamic mean $\langle \ldots \rangle$ is taken with respect to the Hamiltonian (4.11). In order that the NSW exist as elementary excitations, the spectral function $\mathscr{P}_k(\omega)$ must have a narrow peak at the frequency $\omega = \Omega_{nk}$ the width of which $\Delta\omega_{nk}$ must be small compared with the band width of the nuclear spin-wave spectrum $\delta\omega_n$.

For the temperature region $\kappa T \gg \hbar\omega_n$, the method of moments (see, for example, Abragam, 1963) can be used: the first moment

$$\bar{\omega} = \int \mathscr{P}_k(\omega)\, \omega \, d\omega \, \bigg/ \int \mathscr{P}_k(\omega)\, d\omega$$

gives the position of an individual level in the NSW band, while the second moment $\overline{(\Delta\omega)^2} = \overline{(\omega - \bar{\omega})^2} = \overline{\omega^2} - (\bar{\omega})^2$ is the width of this level.

Direct calculations (De Gennes et al., 1963) convince us that the first moment indeed coincides with the NSW frequency (4.13): $\bar{\omega} = \Omega_{nk}$. As regards the second moment, it can be found in the form of an expansion in the inverse temperature $1/T$. Retaining the two first terms, the calculation leads to the following expression for the mean square of the NSW level width (Pincus, 1963):

$$\Delta\omega_{nk} \simeq (\Delta\omega_{nk})_{k=0} = (\Delta\omega_n)_\infty \left(1 - \frac{3}{8} \frac{|\delta\omega_n|}{\omega_n} \right), \tag{4.18}$$

where

$$(\Delta\omega_n)_\infty = \sqrt{\frac{I(I+1)}{24\pi S^2}} \, \frac{\omega_{n0}^2}{\omega_E^{3/4} \omega_e^{1/4}} \tag{4.19}$$

is the mean square width of the level at high temperatures $(T \to \infty)$.

Evaluating the magnitude of the ratio

$$\frac{\delta\omega_n}{\Delta\omega_{nk}} \simeq [24\pi/I(I+1)]^{1/2} (\omega_E/\omega_e)^{3/4} I_z, \tag{4.20}$$

it is possible to find some critical temperature T_0, below which this ratio is large. Since

$$I_z \simeq \frac{\hbar\omega_{n0} I(I+1)}{3\kappa T}, \tag{4.21}$$

we have $|\delta\omega_n|/\Delta\omega_{nk} \gg 1$ for temperatures $T \gg T_0$, where

$$T_0 = \sqrt{\frac{8\pi}{3} I(I+1)} \left(\frac{\omega_E}{\omega_e} \right)^{3/4} \frac{\hbar\omega_{n0}}{\kappa}. \tag{4.22}$$

Since always $\omega_E \gg \omega_e$, we have $T_0 \gg \hbar\omega_{n0}/\kappa$ and thus there exists a temperature interval

$$T_0 \gg T \gg \hbar\omega_{n0}/\kappa \tag{4.23}$$

in which, on the one hand, the nuclear spin-system can be considered as paramagnetic with a small degree of polarization and, on the other, the condition for the existence of collective spin vibrations—nuclear spin waves—is fulfilled. Strictly speaking, all this refers to the case when the Suhl–Nakamura interaction is the basic cause for the broadening of the NSW level.

For $\omega_E/\omega_e \sim 10^3$, $I = 5/2$ and $\omega_{n0} \sim 10^9$ sec^{-1} we obtain from (4.22) that $T_0 \sim 10°K$. Consequently, condition (4.23) is fulfilled at temperatures $T \sim 0.1 - 1°K$.

Let us now turn our attention to the second term in expression (4.18) for the mean square width of the NSW level. It is proportional to the dynamic shift of the NMR frequency and is caused by correlation effects. Owing to the correlation in the movements of the nuclear spins, the second moment of the line is somewhat reduced at low temperatures in comparison with its value (4.19) at high temperatures. However, as can be seen from expression (2.60) (see Chapter II, Sec. 3) and from the estimates presented in the case of a ferromagnet, this decrease is very insignificant, so that it can be neglected.

If the width of NSW levels is ignored, their frequency spectrum can be obtained from the classical equations of motion of the form (2.52) as one of the branches of the spectrum of coupled electron and nuclear spin vibrations. For this purpose the nonhomogeneous part of the exchange energy of a ferromagnet $(Ja^2/M_0^2)(\nabla M)^2$ (see, for example, Akhiezer et al., 1967) should be taken into account in the initial expression for the energy (2.51). The solution of the equation replacing (2.53) will then have the form of plane waves: $M^\pm, m^\pm \sim e^{i\omega t + ikr}$ and we will obtain instead of (2.58) and (2.59) the expressions

$$\Omega_{ek} = \omega_k + \gamma_e A_0 m$$

$$\Omega_{nk} = \omega_{n0}\left(1 - \frac{\gamma_e A_0 m}{\omega_k}\right) - \gamma_n H, \tag{4.24}$$

respectively, the second of which coincides with (4.13).

3. Indirect Interaction Between Spin Waves and Nuclear Spin Waves in Antiferromagnets

Nakamura (1958) calculated the indirect interaction between nuclear spins via spin waves for a uniaxial antiferromagnet with the antiferromagnetic axis directed along the main axis of the crystal. As has already been said before, in this case the spin waves have a large energy gap of the order of $\gamma_e \hbar \sqrt{H_E H_a}$. On account of the condition

$\omega_e \gg \omega_{n0}$, ω_T we can use here the same approximation of perturbation theory as in the previous case of a ferromagnet. Accordingly, the following expression is found for the effective Hamiltonian of the indirect interaction:

$$\mathscr{H}_{eff} = \frac{1}{2} F \left[\sum_j (I_j^z)^2 + \sum_l (I_l^z)^2 \right] - \frac{1}{2} \sum_{j > j'} B_{jj'} (I_j^+ I_{j'}^- + I_j^- I_{j'}^+) -$$

$$- \frac{1}{2} \sum_{l > l'} B_{ll'} (I_l^+ I_{l'}^- + I_l^- I_{l'}^+) - \sum_{jl} C_{jl} (I_j^+ I_l^- + I_j^- I_l^+), \qquad (4.25)$$

where

$$F = \frac{A^2 S}{\hbar N} \sum_k \omega_k^{-1}, \qquad B_{jj'} = \frac{A^2 S}{\hbar N} \sum_k \frac{\cos (k R_{jj'})}{\omega_k},$$

$$C_{jl} = \frac{A^2 S}{\hbar N} \sum_k \frac{\gamma_k \cos (k R_{jl})}{\omega_k}, \qquad \gamma_k = \frac{1}{z} \sum_R e^{ikR}.$$

Here the indices of summation j and l in (4.25) refer to the sites of the first and second magnetic sublattice respectively; R is the index of summation over the nearest-neighbor sites of one sublattice for each atom of the other; z is the number of such neighbors. Finally,

$$\omega_k = |\gamma_e| \sqrt{(H_E/2)^2 (1 - \gamma_k^2) + H_E H_a} \qquad (4.26)$$

is the frequency of spin waves with wave vector k (compare with expression (2.42), Chapter II.) All these expressions are, strictly speaking, valid for $H = 0$, when the frequencies of both spin wave branches coincide; they are, however, approximately valid even for $H \neq 0$, provided $H \ll H_{Ea} = \sqrt{H_E H_a}$.

Using the effective Hamiltonian (4.25) of the indirect interaction between nuclear spins, the NSW spectrum of an antiferromagnet can be calculated in a way analogous to the case of a ferromagnet. Correspondingly, the width of the dispersion band will again equal the dynamic NMR frequency shift, while the wavelengths for which spatial dispersion is considerable will be bounded from below by the correlation radius $b_0 \sim a \sqrt{H_E/H_a}$.

However, as we have already seen for the example of the homogeneous resonance frequency in Chapter II, Sec. 3, the correlation effects associated with HFI are most pronounced for an antiferromagnet of the "easy plane" type (and in general for weakly anisotropic antiferromagnets). Therefore, we shall consider in detail the problem of NSW spectrum for this type of antiferromagnet only. For such antiferromagnets, however, the coupling between the vibrations of the electron and nuclear spins at low temperatures (when $\omega_T \sim \omega_e$) becomes so strong that it is impossible to

apply perturbation theory to the indirect interaction between nuclear spins via unperturbed spin waves.

There remains only the second way for the calculation of the NSW spectrum, which was mentioned in the case of ferromagnets. It is necessary to calculate the eigenfrequencies of the coupled vibrations of electron and nuclear spins taking into account spatial dispersion (i.e., the dependence of the exchange part of the energy of an antiferromagnet on the gradient of the sublattice magnetizations M_p) from the classical equations of motion of the form (2.52) (Turov and Kuleev, 1965).

Repeating the calculation which was carried out in Chapter II, Sec. 3, and taking into account the nonhomogeneous part of the exchange energy in (2.63), we obtain again an expression of the type (2.67) for the eigenfrequencies of the coupled vibrations, in which now, however, instead of the homogeneous AFMR frequency ω_e appears the corresponding antiferromagnetic spin wave frequency ω_k. This frequency can be represented for all cases of interest in the form (see, for example, Turov, 1963)

$$\omega_k^2 = \omega_e^2 + \omega_E^2 (ak)^2 , \tag{4.27}$$

where ω_e is the former homogeneous AFMR frequency, which is defined according to expressions (2.79), (2.83) and (2.86) for antiferromagnets of the $CsMnF_3$, $MnCO_3$ and $KMnF_3$ type; ω_E is the exchange interaction constant (which in the nearest neighbor approximation differs from $|\gamma_e|H_E$ by some numerical factor, dependent on the type of lattice; for example, for a simple cubic lattice $\omega_E = |\gamma_e|(H_E/\sqrt{12})$.

Substituting $\omega_e \to \omega_k$ in equation (2.67), it turns out that both coupled vibration frequencies determined by it (the "electron-like" Ω_{ek} as well as the "nuclear-like" Ω_{nk}) have become dependent on the wave vector k. As before, their product remains the same as for the unshifted vibration frequencies

$$\Omega_{nk}\Omega_{ek} = \omega_{n0}\omega_k . \tag{4.28}$$

The approximate expression for the frequencies of the nuclear-like branch corresponding to expression (2.78) (in the absence of dispersion) takes the form

$$\Omega_{nk} \simeq \omega_{n0} \left(\frac{\omega_e^2 + \omega_E^2 (ak)^2}{\omega_e^2 + \omega_T^2 + \omega_E^2 (ak)^2} \right)^{1/2} . \tag{4.29}$$

This is also the NSW frequency sought for the antiferromagnets under consideration, in analogy with the frequency (4.24) for the case of ferromagnets. A more general expression for Ω_{nk} is given by expression (2.74) with the "minus" sign, when ω_e^2 in it is replaced by ω_k^2.

The width of the NSW band, here too, is equal to the difference between the unshifted frequency (ω_{n0}) and the shifted homogeneous NMR frequency ($\Omega_n \equiv \Omega_{nk}$ for $k = 0$)

$$\delta\omega_n = \omega_{n0} - \Omega_n \simeq \omega_{n0} \left(1 - \frac{\omega_e}{\sqrt{\omega_e^2 + \omega_T^2}} \right) . \tag{4.30}$$

With the increase of wave vector k the dispersion of NSW rapidly becomes negligible and its frequency Ω_{nk} approximates the unshifted value of the NMR frequency. The qualitative picture of the NSW spectrum will again be analogous to that which is presented in Figure 7. However, the relative width of the dispersion band

$$\delta\omega_n/\omega_{n0} \simeq 1 - \frac{\omega_e}{\sqrt{\omega_e^2 + \omega_T^2}}$$

and the limiting magnitude of the wave vector k_0, up to which the dispersion is still significant, may be considerably larger than for a ferromagnet.

The limiting magnitude of the vector $k = k_0$, above which there is no dispersion, can be taken at the middle of the dispersion band defined by the equation

$$\Omega_{nk} = \tfrac{1}{2}(\Omega_n + \omega_{n0}). \tag{4.31}$$

From this equation, taking into account (4.29), we find approximately

$$(ak_0)^2 = \left(\frac{\omega_e}{\omega_E}\right)^2 \frac{(\Omega_e/\omega_e + 1)^2 - 4}{4 - (\omega_e/\Omega_e + 1)^2}, \tag{4.32}$$

where Ω_e and ω_e are the shifted and the unshifted AFMR frequencies, respectively.

In the limiting cases of high and low temperatures we have

$$ak_0 \simeq \frac{\omega_e}{\omega_E}\left(1 + \frac{3}{8}\frac{\omega_T^2}{\omega_e^2}\right)$$

for $\omega_T^2 \ll \omega_e^2$, $\tag{4.33}$

$$ak_0 \simeq \frac{\omega_T}{\omega_E}\left(1 + \frac{4}{3}\frac{\omega_e}{\omega_T}\right)$$

for $\omega_T^2 \gg \omega_e^2$.

Consequently, at any temperature, we have to orders of magnitude*

$$k_0 \sim \frac{1}{a}\frac{\Omega_e}{\omega_E}. \tag{4.34}$$

The inverse quantity

$$b_0 = k_0^{-1} \sim a\,(\omega_E/\Omega_e) \tag{4.35}$$

---

* For antiferromagnets all the expressions look simpler if the squares of the corresponding quantities are considered as variables. Thus the width of the dispersion band for the squared frequency is equal to $\delta\,(\omega_n^2) \equiv \omega_{n0}^2 - \Omega_n^2 \simeq \omega_{n0}^2 \dfrac{\omega_T^2}{\Omega_e^2}$, while the value of the wave vector in the middle of this band, at any temperature, is determined by the simple expression $ak_0 = \Omega_e/\omega_E$.

gives, as in the case of a ferromagnet, the effective linear dimensions of the region inside which the movements of nuclear spins are correlated. Note that the correlation radius b_0 in this case is by far larger than the analogous quantity for ferromagnets (4.15), for which the ratio ω_E/Ω_e is raised to the power of 1/2. For antiferromagnets possessing a low-frequency AFMR branch, b_0 constitutes hundreds or thousands of interatomic distances.

Of course, it is meaningful to speak of "nuclear spin waves" as of some elementary excitations with a definite wave vector k only in the case when the level width $\Delta\Omega_{nk}$ corresponding to this state is small in comparison with the width of the dispersion band $\delta\omega_n$. If we assume that, just as in the case of ferromagnets, $\Delta\Omega_{nk}$ is comparable in order of magnitude with the NMR line width, the condition mentioned is fulfilled in this case, not only for helium temperatures but apparently also for hydrogen temperatures. For instance, at 4.2°K the experimental value of $\delta\omega_n$ for the anti-ferromagnetic manganese compounds mentioned above attains ten or more percent of the unshifted frequency ω_{n0}, possibly exceeding by two–three orders of magnitude the NMR line width.

The velocity $v = \partial\Omega_{nk}/\partial k$ and the mean free path $l = v/\Delta\Omega_{nk}$ of the nuclear spin waves can be estimated. According to (4.29) we have

$$v = \frac{\omega_T^2\omega_{n0}^2\omega_E^2 a^2 k}{\Omega_{nk}(\omega_e^2 + \omega_T^2 + \omega_E^2 a^2 k^2)}.$$

In particular, for $k \sim k_0$ in the region of high temperature, where according to (4.33) $ak_0 \sim \omega_e/\omega_E$, we obtain taking (4.13) into consideration

$$v \sim a\omega_T^2\omega_E\omega_{n0}/4\omega_e^2.$$

Taking as an estimate $a = 3 \cdot 10^{-8}$ cm, $\omega_E = 10^{13}$, $\omega_e = 3 \cdot 10^{10}$, $\omega_{n0} = 3 \cdot 10^9$, $\omega_T = 10^{10}$ and $\Delta\Omega_{nk} = 10^6$ sec^{-1}, we find

$$v \sim 10^3 \text{ cm/sec}, \quad l \sim 10^{-3} \text{ cm}.$$

These values for the NSW velocity and free path increase considerably with the lowering of temperature (Turov and Kuleev, 1965).

It is possible that in thin antiferromagnetic plates at low temperatures NSW resonance (the excitation of standing NSW by a homogeneous r.f. field) can be observed, analogous to ferromagnetic spin wave resonance observed in ferro-magnetic films (Seavey and Tannenwald, 1958). Assuming that an integer multiple of half-waves fit in the plate thickness d (for example, in a case when the boundary conditions are such that the nuclear spins are fixed at the surface of the plate) we have

$$k = k_z = \pi p/d, \tag{4.35a}$$

where p is an integer; for the given boundary conditions only harmonics with odd p can be excited (disregarding the homogeneous resonance corresponding to $p = 0$).

Thus, according to (4.29) and (4.35a), a discrete set of resonance peaks should be observed, separated from the frequency of the homogeneous resonance by the relative magnitude (for small p)

$$\delta_p = \frac{\Omega_{np} - \Omega_n}{\Omega_n} \simeq \frac{1}{2} \frac{\omega_E^2 \omega_T^2}{\omega_e^2 (\omega_e^2 + \omega_T^2)} \left(\frac{\pi a}{d} p \right)^2. \tag{4.36}$$

A permissible value is obtained for δ_p in the region of low temperatures. For example, for $\omega_T \sim \omega_e \sim 10^{-3} \omega_E$, $p = 3$, $a = 3 \cdot 10^{-8}$ cm and $d \sim 10^{-3}$ cm we have $\delta_p \sim 10^{-2}$ which in any case exceeds the line width by an order of magnitude.

From an expression analogous to (2.80) (replacing ω_H^2 by ω_k^2), a more general expression can be obtained, determining the magnitude of the resonance field for each harmonic p of standing spin waves at the given frequency ω,

$$\omega_H^2(p) = \omega^2 \left(1 - \frac{\omega_T^2}{\omega^2 - \omega_{n0}^2} \right) - \omega_E^2 \left(\frac{\pi p a}{d} \right)^2. \tag{4.37}$$

For $\omega^2 < \omega_{n0}^2$ this expression defines a set of resonance fields for the nuclear-like branch of spin waves, and for $\omega^2 > \omega_T^2 + \omega_{n0}^2$, a similar set for the electron-like spin waves.

Finally, let us point out that the values of the NSW frequencies fill, in a quasi-continuous way, the whole interval between the shifted (Ω_n) and the unshifted (ω_{n0}) homogeneous NMR frequencies. This circumstance is possibly associated with the effect mentioned in Chapter II, Sec. 3, observable in $KMnF_3$ (and other anti-ferromagnets of this type): as we have indicated, the saturation of the nuclear spin-system takes place even at frequencies ω very far removed from the NMR frequency ω_n corresponding to the given temperature, provided ω falls into the relevant frequency interval between Ω_n and ω_{n0} (Witt and Portis, 1964b). It is possible that the saturation occurs precisely as a result of the excitation of corresponding nuclear spin waves, while lattice defects and twinning planes, which in the theory of Witt and Portis (1964b) are considered as "nucleation centers of saturation" (sites where the spins are fixed), play the role of surfaces providing the necessary boundary conditions for the excitation of nuclear spin waves.

When inhomogeneities are present in the crystal lattice, the excitation of NSW can also lead to an apparent broadening of the NMR line (which, evidently, will increase with the lowering of the temperature) in the usual linear regime at a low level of r.f. power.

CHAPTER V

Relaxation Processes and the NMR Line Width

1. General Remarks

In magnetic crystals, there exist, of course, all those mechanisms of relaxation and NMR line broadening which are responsible for the observed resonance properties of nonmagnetic solids of the corresponding type (metals, analogous nonmagnetic compounds, etc.). Here, however, we shall mainly consider features associated with the presence of magnetic order in the electron spin-system of ferro- and anti-ferromagnets and with its destruction caused by thermal motion.* These features are a consequence of peculiarities in the properties of the electron system itself.

One of the main peculiarities of this kind is the existence of collective spin excitations, describable (at low temperatures) in terms of spin waves.

Interacting with nuclear spins as a result of hyperfine coupling, dipole–dipole and other forces, the spin waves (magnons—in particle language) can be scattered by nuclear spins, as well as absorbed and emitted by them; finally, spin waves can be converted into lattice vibrations—phonons.

The active role of spin waves in relaxation processes can be caused by the following two types of effects.

1) When scattered by a nuclear spin (alone or with the participation of phonons), spin waves create a fluctuating local magnetic field at the nucleus. The latter can induce quantum transitions between nuclear quasi-Zeeman levels.

2) The emission of a spin wave by one nucleus and its absorption by another leads to an indirect (Suhl–Nakamura) interaction between nuclear spins (see Chapter IV).

---

* General treatment of relaxation processes in spin-systems and of the form of the NMR line and its width can be found in a number of books and reviews (see, for example, Abragam, 1963 and Slichter, 1967).

This Suhl–Nakamura interaction can serve as one of the main reasons for the broadening of the NMR line in magnetic crystals.

There is a significant difference between these two types of effects. The first can take place with an isolated nuclear spin in a ferromagnet (antiferromagnet). Therefore, it must be taken into account for any small concentration of magnetic nuclei. The effects of the second type play a role only for a sufficiently high concentration of magnetic nuclei and, thus, they are particularly large when the isotopic composition of the nuclei under consideration is entirely magnetic.

The discussion below will refer to the case of simple ferromagnets and antiferromagnets describable respectively by one and two magnetic sublattices. Nonetheless, in the majority of cases the results will be applicable also to more complex colinear magnetic structures (e.g., ferrimagnets). This is connected with the fact that the optical branches of the spin waves, which appear when the number of magnetic sublattices is large, cannot usually play a significant role in the relaxation processes for NMR. At the same time it must be borne in mind that the energy of the acoustic spin waves for these more complex structures will be determined by some effective parameters, which are characteristic of all magnetic sublattices and of the interaction between them (see, for example, expression (2.21)).

We shall first consider effects of the second type, associated with the Suhl–Nakamura interaction between nuclear spins.

2. The Suhl–Nakamura Broadening

The effective Hamiltonian of the indirect interaction between nuclear spins via spin waves for the case of a ferromagnet was derived in Chapter IV, Sec. 1, and presented in the form (4.8). This expression has the form of one of the so-called secular terms of the pseudodipole Hamiltonian (see the term of type B in the book of Abragam, 1963).

In recent NMR theories, interaction in a system of many spins is usually taken into account by the method of moments developed by Van Vleck (1948). The moments M_n of a resonance line at the frequency ω_0 ($n = 2, 4, 6, \ldots$) can characterize its shape and width.* If the line shape is close to Gaussian, the second moment of this line gives its half-width at the half-height of the resonance peak:

$$\Gamma \simeq \sqrt{M_2} \equiv \overline{[(\omega - \omega_0)^2]}^{1/2}, \tag{5.1}$$

---

* By definition, $M_n = \displaystyle\int_{-\infty}^{\infty} (\omega - \omega_0)^n f(\omega)\, d\omega$, where ω_0 is the resonance frequency and $f(\omega)$ is the line shape function.

while the fourth moment $M_4 = \overline{(\omega - \omega_0)^4}$ is connected with M_2 by the simple relation

$$M_4/M_2^2 \simeq 3.$$

At the same time, if calculation shows that $M_4/M_2^2 \gg 1$, the resonance line can approximately be described by a Lorentzian curve. Accordingly, the half-width of the line

$$\Gamma \simeq \sqrt{M_2} \left(\frac{M_2^2}{M_4}\right)^{1/2} \tag{5.2}$$

is appreciably less than the root-mean-square width $\sqrt{M_2}$ (the second moment).

Thus, to determine the line shape and line width caused by the Suhl–Nakamura interaction (4.8), it is necessary to take into account at least the second and the fourth moments. The second moment for a ferromagnet was found by Suhl (1958):

$$\sqrt{M_2} = \left[\frac{1}{3} I (I + 1) \sum_{j \neq j'} U_{jj'}^2\right]^{1/2} = \left[\frac{I (I + 1)}{24\pi S^2}\right]^{1/2} \frac{\omega_{n0}^2}{\omega_E^{3/4} \omega_e^{1/4}}. \tag{5.3}$$

The evaluation of the fourth moment, carried out by Vatova (1965), led to $M_4/M_2^2 \simeq 6$. Consequently, the line can approximately be considered Gaussian and its half-width is characterized by the second moment (5.3). According to Houston and Heeger (1966), the compound Mn_3O_4 (a ferrite with the hausmannite structure) is a good example of a ferromagnet in which the Mn^{55} NMR line width can be fully explained by the Suhl–Nakamura mechanism.

In an analogous way the mean square half-width of the NMR line of an antiferromagnet (of the EA type) can be calculated using the effective Hamiltonian (4.25). In this calculation, the terms with C_{jl} in (4.25), describing the indirect coupling between the spins of nuclei which belong to different magnetic sublattices, must not be taken into account, since these terms do not commute with the quasi-Zeeman Hamiltonian

$$\mathcal{H}_z = AS \left(\sum_j I_j^z - \sum_l I_l^z\right), \tag{5.4}$$

determining the NMR frequency.* As a result, if we take the approximate expression (4.27) for the energy of spin waves, we obtain (Nakamura, 1958)

$$\sqrt{M_2} \simeq \left[\frac{I (I + 1)}{96\pi S^2}\right]^{1/2} \frac{\omega_{n0}^2 (|\gamma_e| H_E)}{\omega_E^{3/2} \omega_e^{1/2}}. \tag{5.5}$$

---

* According to the theory of moments (see, for example, Abragam, 1963) only the terms conserving the total spin of the nuclei, resonating at a particular frequency, have to be retained in the Hamiltonian of the spin interaction in the calculation of moments which characterize the width and the shape of the corresponding line. Let us remember that in the case of an antiferromagnet, one sublattice resonates at the frequency $\omega \simeq \omega_{n0} = AS/\hbar$, while the other at the frequency $\omega \simeq -\omega_{n0} = -AS/\hbar$.

In particular, for a simple cubic lattice in the nearest neighbor approximation $(|\gamma_e| H_E = \omega_E \sqrt{12})$ instead of (5.5) we have

$$\sqrt{M_2} \simeq \left[\frac{I(I+1)}{8\pi S^2} \right]^{1/2} \frac{\omega_{n0}^2}{(\omega_e \omega_E)^{1/2}}. \tag{5.6}$$

The expression (5.5) was obtained for an anti-ferromagnet of the EA type, for which $\omega_e \simeq |\gamma_e| \sqrt{H_E H_a}$. For this case $\sqrt{M_2}$, or anyway its particular form (5.6), coincides in fact (within a numerical factor) with the corresponding expression (5.3) for a ferromagnet, if we assume for the latter $H = 0$ and, consequently, $\omega_e = |\gamma_e| H_a = \omega_a$.[*]

However, expression (5.5) is, with some reservations, applicable also to other antiferromagnets (for example, of the EP type) provided we interpret ω_e as the corresponding AFMR frequency. It must be borne in mind that this expression was obtained without taking into account correlation effects (see Chapter IV). Therefore, the situation becomes more complex when the NMR frequency is "mixed" with the low frequency branch of the AFMR, in the region of very low temperatures and sufficiently small fields H, where the dynamic coupling and the corresponding frequency shift are large, and expression (5.5), strictly speaking, can no longer be employed. For this case, it is necessary to develop the theory further.

The Suhl–Nakamura mechanism of line broadening gives the necessary order of magnitude for a number of antiferromagnets with a high isotope content of magnetic nuclei. MnF_2 is a typical example of an antiferromagnet of the EA type, for which, apparently, the line width of both the F^{19} and the Mn^{55} nuclei is associated with the Suhl–Nakamura mechanism (see Review, 1965a).

For antiferromagnets of the EP type and other antiferromagnets having low-frequency AFMR branches $(\omega_e \lesssim 10^{11} \text{ sec}^{-1})$, it follows from (5.6) for $\omega_E \sim 10^{13}$ sec^{-1} and $\omega_n = 4 \cdot 10^9 \text{ sec}^{-1}$ that $\Gamma_n \simeq \sqrt{M_2} \sim 4 \cdot 10^6 \text{ sec}^{-1}$. This agrees in order of magnitude with observed values for the half-width of the Mn^{55} line in $MnCO_3$ (Shaltiel, 1966), $KMnF_3$ (Nakamura et al., 1964) and $RbMnF_3$ (Heeger and Teaney, 1964). In the latter case of $RbMnF_3$, the minimal experimental value of the line width, observed at sufficiently large fields H (for which correlation effects can be neglected), turned out to be an order of magnitude less than the theoretical value presented above.

However, it should not be forgotten in a detailed quantitative comparison of theory with experiment that, first, the NMR line does not have a strict Gaussian shape and therefore the second moment $\sqrt{M_2}$ is not exactly equal to the half-width of the line. Secondly, in the case of antiferromagnets, the more general expression (5.5) containing two exchange parameters must be used.

---

[*] In the general case, expression (5.5) contains two exchange parameters $|\gamma_e| H_E$ and ω_E which can be expressed one in terms of the other only on condition that the exchange interaction between spins belonging to the same magnetic sublattice can be neglected.

3. Nonuniform Broadening

A trivial reason for the finite NMR line width in real crystals can be found in the nonuniformity of the local magnetic field, acting on the nuclear spins. It can be associated with nonuniformities in the distribution of magnetization (with respect to direction as well as to magnitude) leading to nonuniformities in the dipole and hyperfine fields.

This reason is particularly important for NMR involving nuclei in domain walls (see Chapter VII, Sec. 3). For nuclear spins $I > \frac{1}{2}$ the quadrupole effects must also be taken into account (see Chapter VI, Sec. 2).

A second possible mechanism of nonuniform broadening of the NMR line can be brought about by the dynamic NMR frequency shift considered in Chapter II, as a result of the coupling of the electron and nuclear spin vibrations at low temperatures. If the line width of the electron magnetic resonance has a nonuniform origin (and at low temperatures the share of nonuniform broadening predominates), this leads to a nonuniform shift in the NMR frequency, i.e., to line broadening, as the NMR frequency depends on the frequency of the electron resonance.

It is not difficult to derive from expressions (2.59) and (2.78) that the nonuniform broadening $\Delta\omega_e$ of the electron resonance must lead to the following values for the NMR line width:

$$\left(\frac{\Delta\omega_n}{\omega_n}\right)_{nonunif} = \frac{\gamma_e A_0 m_0}{\omega_e}\left(\frac{\Delta\omega_e}{\omega_e}\right)_{nonunif} \qquad \text{(ferromagnet)}, \qquad (5.7)$$

$$\left(\frac{\Delta\omega_n}{\omega_n}\right)_{nonunif} = \frac{\omega_T^2}{\omega_e^2 + \omega_T^2}\left(\frac{\Delta\omega_e}{\omega_e}\right)_{nonunif} \qquad \text{(antiferromagnet)}. \qquad (5.8)$$

It should be pointed out that for antiferromagnets of the EP type the nonuniform broadening (5.8) of the NMR line can play a particularly important role at low temperatures and small fields, when $\omega_T \sim \omega_e$ (remember that $\omega_T^2 = \gamma_e^2 A_0 m_0 H_E \sim 1/T$).

In such antiferromagnets it is difficult to get rid of the large value of $(\Delta\omega_e/\omega_e)_{nonunif}$ because of the strong influence of various internal nonuniform stresses on the antiferromagnetic resonance frequency (Borovik-Romanov and Rudashevskii, 1964; Turov and Shavrov, 1965). These stresses, as a result of magnetostriction, produce a nonuniform effective anisotropy field H_{ms}, which, although small by itself, can be appreciable in its influence on the antiferromagnetic resonance frequency ω_e, because it enters the expression for ω_e^2 in combination with the exchange field: $\gamma_e^2 H_{ms} H_E$.[*] As a result, the quantity H_A, entering the expression for the AFMR

---

[*] Note that the field H_{ms} is isotropic in the first approximation in the sense that it does not depend on the direction of the antiferromagnetic vector in the easy plane.

frequency (see, for example, (2.79) or (2.83)), becomes nonuniform, which leads to a spread in the frequencies ω_e and, consequently, Ω_n. It is clear that the role of this effect depends on the magnitude of the constant field H; it is important only when the field-dependent term in ω_e^2 is not very large in comparison with $\gamma_e^2 H_E H_{ms}$.

This mechanism can explain the very strong dependence of the NMR line width on the field, observed at low temperature in $MnCO_3$ (the sharp increase of $\Delta\omega$ with decreasing H), if it is assumed that for a small field this nonuniform mechanism is the basic cause of broadening (Shaltiel, 1966). This is seen in Figure 11 where the experimental values and the theoretical curve for the dependence of the line width on the field are shown for $MnCO_3$. The curve was calculated on the assumption that nonuniform broadening is caused by a spread in the values of H_Δ in an interval of 1.2 Oe. At high fields H, the dynamic shift, together with the nonuniform broadening, become insignificant and, as has already been mentioned before (see Sec. 2), the line width can be explained by the Suhl–Nakamura mechanism.

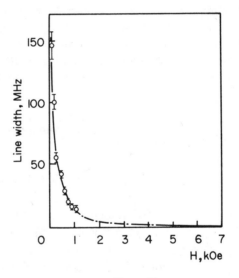

Figure 11

The Mn^{55} NMR line width in $MnCO_3$ as a function of the magnetic field at $4.2°K$ (Shaltiel, 1966).

It is possible that the singularities in the temperature and field dependence of the line width observed in $RbMnF_3$ (Heeger and Teaney, 1964) and in $KMnF_3$ (Nakamura et al., 1964) are associated with this mechanism of nonuniform dynamic NMR frequency shift.

4. Relaxation Times and Correlation Functions

Let us turn now to the discussion of relaxation processes and NMR line widths caused by the fluctuation of the local field at the nucleus. Let us denote by δH the difference between the true (instantaneous) value of the local field H_{loc} and its thermodynamic mean value $\langle H_{loc} \rangle$:

$$\delta H = H_{loc} - \langle H_{loc} \rangle. \tag{5.9}$$

Let Z be the quantization axis for the nuclear spin (which, obviously, coincides with the direction of $\langle H_{loc} \rangle$). It is clear that the roles of the transverse components of the fluctuating field $\delta H^\pm$ and of its longitudinal component δH^z will be altogether different.

If the frequency spectrum of the fluctuations $\delta H^\pm$ (determined by the vibration eigenfrequency spectrum of the system producing δH) contains the NMR frequency $\omega_n = \gamma_n \langle H_{loc}^z \rangle$, this field component will induce quantum transitions between quasi-Zeeman levels of the nucleus, with a change of one in the magnetic quantum number, $\Delta m = \pm 1$. As a result of such transitions, the longitudinal magnetization of the nuclear system can change and approach equilibrium with the "lattice." Thus, $\delta H^\pm$ can be responsible for the longitudinal (or, otherwise, the spin-lattice) relaxation of the nuclei. This relaxation is the cause of the finite lifetime of nuclear spins in the excited state and it can definitely contribute to the NMR line width, leading to a so-called nonsecular broadening (see, for example, Abragam, 1963).

The longitudinal field δH^z (considered as a perturbation) cannot cause transitions with a change in the longitudinal magnetization of the nuclei, since the corresponding term in the Hamiltonian, $-\hbar \gamma_n I^z \delta H^z$, commutes with I^z. However, the spread in the instantaneous values of the resulting longitudinal local magnetic field ($\langle H_{loc}^z \rangle + \delta H^z$) associated with the field δH^z will cause a so-called secular (or adiabatic) broadening of the NMR line. If a macroscopic transverse nuclear magnetization $m_\perp$ is created in the sample by some means (for example, an in-phase precession of all the nuclear spins can be induced by means of a rotating rf field), the fluctuations δH^z will destroy the phase synchronism in the precession of various spins and consequently attenuate the resulting magnetization $m_\perp$ (conserving the longitudinal magnetization m_z). Thus, it is to be expected that δH^z will be one of the causes for the relaxation of the transverse component of the nuclear magnetization.

The detailed theoretical treatment of relaxation and broadening of the NMR line is based on a Hamiltonian of the following form:

$$\mathcal{H} = \mathcal{H}_S - \hbar \omega_n \sum_l I_l^z - \hbar \gamma_n \sum_l I_l \delta H_l, \tag{5.10}$$

where $\mathcal{H}_S$ is the Hamiltonian of the electron system and the lattice (subsystem S), without taking into account the interaction with nuclear spins; the second term is

the quasi-Zeeman energy of the nuclei; finally, the last term constitutes the interaction between the nuclear spins and the fluctuating part of the local fields. The summation is carried out on all the indices l of the magnetic nuclei.

There are several methods in the theory of the dynamic properties of spin systems describable by the Hamiltonian (5.10).* The most widespread approach considers the last term in (5.10) as a small perturbation and thus derives the phenomenological Bloch equation for the nuclear magnetization

$$\frac{d\boldsymbol{m}}{dt} = \gamma_n(\boldsymbol{m} \times \boldsymbol{H}) - \boldsymbol{i}\,\frac{m_x}{T_2} - \boldsymbol{j}\,\frac{m_y}{T_2} - \boldsymbol{k}\,\frac{m_z - m_0}{T_1}, \tag{5.11}$$

in which the transverse (T_2) and the longitudinal (T_1) relaxation times can be expressed via the correlation functions of the subsystem S (electrons and lattice) which produces the fluctuating field δH_l. As starting point there are the equations of motion for the density matrix of the whole system with the Hamiltonian (5.10), which is subsequently averaged over the variables of the subsystem S. Equation (5.11) can be derived from the averaged equation if, first, the correlation time τ_C of the fluctuations δH_l characteristic of subsystem S satisfies a condition of the form

$$\tau_C \ll T_1, T_2$$

and, second, the interaction between the nuclear spins can be neglected (both the direct interaction and the indirect interaction via subsystem S). Accordingly, the relaxation times are determined by the following relations (Moriya, 1956; 1962):

$$\frac{1}{T_1} = \gamma_n^2 \int_{-\infty}^{\infty} f_1(t) \cos \omega_n t\, dt, \tag{5.12}$$

$$\frac{1}{T_2} = \frac{1}{2T_1} + \gamma_n^2 \int_{-\infty}^{\infty} f_0(t)\, dt, \tag{5.13}$$

where $f_1(t)$ and $f_0(t)$ are the correlation functions for the fluctuations of the transverse and the longitudinal components of the local field respectively,

$$f_1(t) = \tfrac{1}{2} \langle \{\delta H_l^+(t)\, \delta H_l^-(0)\} \rangle_S, \tag{5.14}$$

$$f_0(t) = \langle \{\delta H_l^z(t)\, \delta H_l^z(0)\} \rangle_S. \tag{5.15}$$

Here the field δH_l is considered as an operator, while

$$\delta H_l(t) = e^{i\mathscr{H}_{st}/\hbar}\delta H_l e^{-i\mathscr{H}_{st}/\hbar} \tag{5.16}$$

is the Heisenberg representation of this operator with the Hamiltonian of the subsystem S. The averaging is also carried out over the states of the same Hamiltonian

---

* See Abragam (1963), Aleksandrov (1964), and Slichter (1967).

$$\langle 0 \rangle_s = \text{Tr}\,(0 e^{-\beta \mathcal{H}_s})/\text{Tr}\,(e^{-\beta \mathcal{H}_s}) \qquad (\beta = 1/\kappa T). \qquad (5.17)$$

Finally, the braces denote the symmetrized operator product

$$\{AB\} = \tfrac{1}{2}(AB + BA).$$

Thus, in accordance with the above physical considerations, the rate of longitudinal relaxation $1/T_1$ is determined by the Fourier component (for $\omega = \omega_n$) of the time correlation function $f_1(t)$ of the transverse components of δH, while the full rate of transverse relaxation $1/T_2$ (determining the half-width of the Lorentzian line Γ_n) is a sum of two terms. The first of these is expressed through the same correlation function $f_1(t)$ and characterizes the nonadiabatic (nonsecular) contribution to the line width. The second term in (5.13) is determined by the correlation function $f_0(t)$ of the longitudinal components of δH. It characterizes the adiabatic (secular) part of the transverse relaxation rate and the line width.*

A significant feature of this method is that the subsystem S plays only a passive role, as a thermal reservoir and a source of fluctuations of the local field. Its active role, i.e., the fact that the electron spins themselves add a resonant contribution to the rf susceptibility as a result of HFI, is, in fact, ignored. A more general theory of this type should lead, on the basis of the Hamiltonian (5.10), to a system of coupled equations of motion for the electron and nuclear magnetizations. There is no such theory at present. If, however, we restrict ourselves to linear NMR effects, we can attempt to carry out a direct calculation of the full rf susceptibility of the electron–nuclear spin system near the NMR frequency bypassing the equations of motion for the magnetic moments. This can be accomplished using the expressions of the general theory for the linear reaction of the system to an external disturbance, the rf field in our case (see, for example, Tyablikov, 1965). Such an attempt was undertaken in the work of Kurkin, Sokolov and one of the present authors.**

The underlying idea of the calculations is the following. Let the concentration of magnetic nuclei be so small that correlation effects associated with direct and indirect interactions between nuclei can again be neglected. In this case we can consider one nuclear spin interacting with the S-system. Let us write the operator of the total magnetic moment of the system in the form

$$\mathfrak{M} = M + \mu,$$

---

* The terms "nonadiabatic" or "adiabatic" contribution corresponding to the first and the second term in (5.13) are associated with the fact that relaxation processes of the first type are accompanied by transfer of energy from the nuclear spins to the "lattice," while the processes of the second type proceed without any transfer of energy.

** See Turov et al. (1967), Kurkin and Turov (1967), Turov and Kurkin (1968). Another approach in the spirit of the theory of rate equations was developed by Bar'yakhtar et al. (1968).

where $\mu = \hbar\gamma_n I$ is the magnetic moment of the nucleus, while M is the total magnetic moment of the electrons. In the model of localized electrons

$$M = \sum_j \hbar\gamma_j S_j,$$

where $\hbar S_j$ is the spin operator (or the total angular momentum operator) of the atom. The total magnetic susceptibility of such an electron–nuclear system with respect to a uniform rf field can be represented in the form (Tyablikov, 1965)

$$\chi_{\alpha\beta}(\omega) = - \langle \mathfrak{M}_\alpha \,|\, \mathfrak{M}_\beta \rangle_\omega = - \langle M_\alpha \,|\, M_\beta \rangle_\omega -$$

$$- \langle M_\alpha \,|\, \mu^\beta \rangle_\omega - \langle \mu^\alpha \,|\, M_\beta \rangle_\omega - \langle \mu^\alpha \,|\, \mu^\beta \rangle_\omega . \tag{5.18}$$

In this expression we have denoted by $\langle A \,|\, B \rangle_\omega$ the Fourier components of the double-time retarded Green's function, defined by the relations

$$\langle A \,|\, B \rangle_\omega = \int_{-\infty}^{\infty} d(t - t') \langle A(t) \,|\, B(t') \rangle \, e^{i\omega(t-t')}, \tag{5.19}$$

$$\langle A(t) \,|\, B(t') \rangle = - i\theta(t - t') \langle [A(t), B(t')] \rangle, \tag{5.20}$$

where $\theta(t - t') = 0$ or 1 for $t < t'$ or $t > t'$, respectively, $[A, B] = AB - BA$ is the commutator, and $A(t) = e^{i\mathcal{H}t/\hbar} A e^{-i\mathcal{H}t/\hbar}$ is the Heisenberg representation of the operator A with the Hamiltonian

$$\mathcal{H} = \mathcal{H}_S - \hbar\omega_n I^z - \mu^z \delta H^z - \tfrac{1}{2}\mu^+ \delta H^- - \tfrac{1}{2}\mu^- \delta H^+ . \tag{5.21}$$

The mean in (5.20) is determined by an expression of the form (5.17), with the full Hamiltonian $\mathcal{H}$ instead of the Hamiltonian $\mathcal{H}_S$.

In subsequent computations, the perturbation theory is used for the Green's functions in a form proposed by Tyablikov and Bonch-Bruevich (1962) (see Tyablikov, 1965). Again, the interaction between a nuclear spin and the fluctuating field δH^α, i.e., the last three terms in (5.21), is considered as the perturbation.

In the zero-order perturbation ($\delta H^\alpha = 0$), the first term in (5.18) gives the unperturbed electron susceptibility at the frequency ω, the second and the third terms are zero, while the last term is the rf susceptibility of the free nuclear spin: $\chi_{\alpha\beta}^{(n)}(\omega) = - \langle \mu^\alpha \,|\, \mu^\beta \rangle_\omega^0$ (the index 0 on the Green's function indicates that it is calculated for $\delta H^\alpha = 0$). In circular coordinates ($\alpha = z, +, -$) only one of the components of the tensor $\chi_{\alpha\beta}^{(n)}$, namely $\chi_{+-}^{(n)}$, has resonant character and is equal to

$$\chi_{+-}^{(n)} = - \langle \mu^+ \,|\, \mu^- \rangle_\omega^0 = - \frac{2\gamma_n \langle \mu^z \rangle^0}{\omega - \omega_n} . \tag{5.22}$$

In the first order in δH^α, the functions $\langle M_\alpha \,|\, \mu^\beta \rangle_\omega$, $\langle \mu^\alpha \,|\, M_\beta \rangle_\omega$ become nonzero and they acquire a resonant property near the NMR frequency. Starting with second-

order perturbation, corrections which have a resonance at $\omega \sim \omega_n$ are also introduced in the functions $\langle M_\alpha | M_\beta \rangle_\omega$.* Collecting all terms up to and including second order, the part of the full susceptibility which has a resonant character near the NMR frequency can be represented in the form

$$\Delta \chi_{\alpha\beta}(\omega) = - \Lambda_{\alpha\beta} \langle \mu^+ | \mu^- \rangle_\omega . \tag{5.23}$$

Here $\Lambda_{\alpha\beta}$ is a complex tensor which, although it depends on frequency, has no singularity near the NMR frequency and therefore can be taken at $\omega = \omega_n$. In the general case, $\Lambda_{\alpha\beta}$ are the Fourier components of a very complicated multi-time correlation function of the operators M_α and δH^α, and we shall not describe them here, referring the reader to the original work of Turov and Kurkin (1968). In the particular case of a simple ferromagnet in the approximation of noninteracting spin waves and in the presence of axial symmetry in the system (when the quantization axes of the electron and the nuclear spins coincide and are directed along the symmetry axis), the tensor $\Lambda_{\alpha\beta}$ takes the form

$$\Lambda_{\alpha\beta} = (\tfrac{1}{2} A_0 \chi^{(e)}_{+-} - 1)^2 \delta_{\alpha, +} \delta_{\beta, -} , \tag{5.24}$$

where $\chi^{(e)}_{+-}$ is the electron susceptibility (not perturbed by interaction with the nuclei) with respect to the circularly polarized field of frequency $\omega \sim \omega_n$.

Since the function $- \langle\langle \mu^+ | \mu^- \rangle\rangle_\omega$ in expression (5.23) in fact gives the intrinsic rf susceptibility of the nuclei, where the interaction with the fluctuating field of the electrons is taken into account, the factor $\Lambda_{\alpha\beta}$ can be considered as an amplification coefficient associated with the contribution of the electrons to the full resonant susceptibility near the NMR frequency. Because of the complex character of $\Lambda_{\alpha\beta}$ the amplification of the NMR signal is accompanied by mixing of the dispersion and absorption curves. In the particular case of a ferromagnet mentioned above, we simply obtain the result of the phenomenological treatment, presented in Chapter III, Sec. 1.

Expression (5.23) reduces NMR to the calculation of the Green's function $\langle \mu^+ | \mu^- \rangle_\omega$. As a result of the interaction with δH^α, the resonance frequency, being a pole of this Green's function, is shifted in the complex plane (with respect to the pole of the nonperturbed function (5.22)). Thus the fluctuating field δH^α leads to a shift and broadening of the NMR line, and if the nuclear spin $I > \tfrac{1}{2}$, then, in general, splitting will also result. The final expression for the Green's function $\langle \mu^+ | \mu^- \rangle_\omega$ can be written in the form**

---

 * Although the resonance of $\langle\langle M_\alpha | M_\beta \rangle\rangle_\omega$ at the NMR frequency appears in second order, while the functions $\langle\langle M_\alpha | \mu^\beta \rangle\rangle_\omega$ and $\langle\langle \mu^\alpha | M_\beta \rangle\rangle_\omega$ have a resonance in the first order, the main contribution to the NMR susceptibility comes from the former functions, because they contain an additional large multiplier of the order γ_e/γ_n which the other functions do not show.

** The calculation is carried out according to the perturbation theory for the mass operator Σ (Tyablikov, 1965) to second order in δH^α (see works cited in footnote on p. 81).

$$\langle \mu^+ | \mu^- \rangle_\omega = \frac{(\gamma_n \hbar)^2 \omega_n}{\kappa T} (2I + 1)^{-1} \sum_{m=-I}^{+I} \frac{I(I+1) - m(m-1)}{\omega - \omega_n - \Sigma(\omega_n, m)}, \qquad (5.25)$$

where

$$\Sigma = \tfrac{1}{2} {}_+\langle \delta H^z | \delta H^z \rangle_{\omega=0}^{(S)} + \tfrac{1}{4} {}_+\langle \delta H^+ | \delta H^- \rangle_{\omega=\omega_n}^{(S)} +$$

$$+ \tfrac{1}{2}(1 - 2m) \left[\langle \delta H^z | \delta H^z \rangle_{\omega=0}^{(S)} - \tfrac{1}{2} \langle \delta H^+ | \delta H^- \rangle_{\omega=\omega_n}^{(S)} \right]. \qquad (5.26)$$

Here, the second line makes use of the Green's function of the commutator of the corresponding operators introduced by relations (5.19) and (5.20) (the index S means that they are computed for $\mathcal{H} = \mathcal{H}_S$), while the same symbols appearing in the first line, identified by a "plus" sign on the left, define, by means of the same relations (5.19) and (5.20), the Green's functions of the anticommutator. In other words, the functions ${}_+\langle A | B \rangle$ are obtained from the functions $\langle A | B \rangle$ by substituting in (5.20) the anticommutator $AB + BA$ for the commutator $[A, B]$ (the commutator and anticommutator Green's functions). The numerator inside the summation in (5.25) is given in the high temperature approximation $\kappa T \gg \hbar \omega_n$. The summation in (5.25) is carried out over all the possible $2I + 1$ projections of the nuclear spin; the first term (for $m = -I$), however, vanishes identically, and therefore the sum actually contains $2I$ terms.

Thus, expression (5.25) describes in the general case a multiplet of $2I$ resonance lines. The central peak corresponds to $m = \tfrac{1}{2}$, while the $2I - 1$ satellites are arranged symmetrically on both sides. As will be shown in Chapter VI, this splitting is completely analogous to the splitting caused by the quadrupole interaction between the nucleus and the crystal field.

The quantity Σ is complex. The line shift and splitting are determined by the real part of Σ, while its imaginary part describes the broadening of the corresponding multiplet components. Isolating the imaginary part of Σ for the central component ($m = \tfrac{1}{2}$), it is not difficult to show that an expression is obtained which coincides with the right member of (5.13):

$$\Gamma_n = \operatorname{Im} \Sigma (m = \tfrac{1}{2}) = 1/T_2. \qquad (5.27)$$

Thus, the two methods considered give in fact the same result for the NMR line width in the case $I = \tfrac{1}{2}$.* The spin-lattice relaxation time T_1, on the other hand, cannot be obtained in the theory of the linear reaction.** However, if T_1 is defined as the time for the establishment of the equilibrium value of the longitudinal nuclear

---

* It must, however, be borne in mind that if the splitting obtained for $I > \tfrac{1}{2}$ in the second method does not exceed in magnitude the broadening of the individual components, it will be interpreted as some additional broadening of a combined curve resulting from the superposition of all components.

** Although it is not difficult to see that $1/T_1$, defined by (5.12), coincides with the imaginary part of the second term in Σ, (5.26).

magnetization (or as the time for the establishment of the equilibrium population difference), it can be calculated directly from the quantum transitions between the Zeeman sublevels of the nucleus produced by the perturbing field δH^α. For example, for a system of two levels ($I = \frac{1}{2}$)

$$\frac{1}{T_1} = W_\downarrow + W_\uparrow, \tag{5.28}$$

where $W_\downarrow$ and $W_\uparrow$ are the mean transition probabilities (in unit time) from the upper and the lower level, respectively (see, for example, Slichter, 1967). Taking $V = -\gamma_n \hbar I \delta H$ as the perturbation operator and calculating the resulting transition probabilities

$$W_{m \to m', f \to f'} = \frac{2\pi}{\hbar} |(fm|V|f'm')|^2 \, \delta(E_f + E_m - E_{f'} - E_{m'}),$$

where m, m' and f, f' are the initial and the final states of the nuclear spin and the subsystem S, $W_\downarrow$ and $W_\uparrow$ can subsequently be found by averaging these probabilities over the initial states of subsystem S and summing over its final states

$$W_{\uparrow,\downarrow} = \frac{\displaystyle\sum_{f,f'} W_{\pm 1/2 \to \mp 1/2,\, f \to f'} e^{-\beta E_f}}{\displaystyle\sum_f e^{-\beta E_f}}. \tag{5.29}$$

The expression obtained for $1/T_1$ in such calculations coincides with expression (5.12), and thus also with the imaginary part of the second term in (5.26).*

The general expression (5.26) for the complex NMR frequency shift Σ contains more information than (5.13). The imaginary part of Σ describes the width of each component of the NMR line, while the real part, in addition to line splitting, gives the frequency shift, which is identical for all the components. This shift, which is retained also for $I = \frac{1}{2}$, is determined by the first line in (5.26),

$$\delta\omega_n = \text{Re}\,\Sigma(m = \tfrac{1}{2}). \tag{5.30}$$

In the rest of this chapter, we shall limit the discussion to the case $m = \frac{1}{2}$, which corresponds either to an isolated single line for $I = \frac{1}{2}$ or to a central line for $I > \frac{1}{2}$. It follows from the definition of the anticommutator Green's function that the first term in (5.26) is pure imaginary and, consequently, it contributes only to the line width (a contribution identical with the second term in (5.13)). Therefore, the shift (5.30) may be caused only by the Green's function $_+\langle\langle \delta H^+ | \delta H^- \rangle\rangle_{\omega_n}^{(S)}$ (its real part).

---

* In the case $I > \frac{1}{2}$ it is necessary to introduce for each pair of neighboring quasi-Zeeman levels their own spin-lattice relaxation time T_{1m}. It can only be assumed that the rate of spin-lattice relaxation $1/T_{1m}$ for the corresponding m will be determined by the imaginary part of the two Green's functions (the anti-commutator and the commutator functions) of the transverse components of δH (the second and the fourth term in (5.26)).

5. Fluctuations of the Hyperfine and the Dipole Fields of Localized Electrons

The basic contribution to the local field for nuclei of magnetic (and in some cases nonmagnetic) atoms is due to a HFI of the form ASI. The resulting fluctuations in the local field are expressed through the electron spin fluctuations*

$$\delta H_j = - (A/\gamma_n \hbar) \delta S_j, \tag{5.31}$$

where $\delta S_j = S_j - \langle S_j \rangle$. Thus the relaxation times, in this case, are expressed through the spin correlation functions

$$f_1(t) = \tfrac{1}{2} (A/\gamma_n \hbar)^2 \langle \{\delta S_j^+ (t) \, \delta S_j^- (0)\} \rangle_s, \tag{5.32}$$

$$f_0(t) = (A/\gamma_n \hbar)^2 \langle \{\delta S_j^z(t) \, \delta S_j^z(0)\} \rangle_s. \tag{5.33}$$

The correlation functions have been calculated by various methods for three temperature regions: 1) for high temperatures, appreciably higher than the Curie (Néel) point, $T \gg T_{C,N}$; 2) near the transition temperature, above $T_{C,N}$, and 3) for low temperatures, $T \ll T_{C,N}$, in the region of applicability of the spin wave theory.**

High Temperatures

The explicit form of the correlation functions f_0 and f_1 depends on the frequency spectrum of the local field fluctuations. In the paramagnetic region, for $T \gg T_{C,N}$, it is usually assumed that the fluctuation spectrum of δH follows a Gaussian distribution. Accordingly, the following relations were obtained for $f_0(t)$ and $f_1(t)$ (Moriya, 1956):

$$f_0(t) = f_1(t) = \tfrac{1}{3} S(S+1)(A/\gamma_n \hbar)^2 \exp\left(-\tfrac{1}{2} \omega_E^2 t^2\right), \tag{5.34}$$

where $\omega_E^2 = \dfrac{2}{3} \dfrac{\mathscr{I}^2}{\hbar^2} ZS(S+1)$; $\mathscr{I}$ is the exchange integral for the nearest neighbor magnetic atoms, Z is the number of nearest neighbors. In this case expressions (5.12)–(5.15), taking (5.31)–(5.34) into account, give

$$\frac{1}{T_1} = \frac{1}{T_2} = \frac{\sqrt{2\pi} S(S+1) A^2}{3\hbar^2 \omega_E} \sim \omega_{n0} \frac{\omega_{n0}}{\omega_E}. \tag{5.35}$$

(Here $\omega_{n0} \sim AS/\hbar$ is the NMR frequency at low temperatures). The equality $T_1 = T_2$

---

* If the orbital momentum is not locked, it is necessary to treat S as the total angular momentum of the atom (ion) J.

** See Moriya (1956), Van Kranendonk and Bloom (1956), Mitchel (1957), Moriya (1962).

is the important feature in expression (5.35). An exact equality is obtained as a result of the assumption of isotropic HFI. If HFI is not isotropic, it can only be asserted that $T_1 \sim T_2$.

The result presented in (5.35) is valid both for ferromagnets and for antiferromagnets. For $\omega_{n0} = 4 \cdot 10^9$ sec^{-1} and $\omega_E = 10^{13}$ sec^{-1} this expression gives $T_1 \sim T_2 \sim 10^{-6}$ sec, which in a number of cases agrees in order of magnitude with experiment (see Review, 1965a, and also Wolker, 1966).

Also note that a relation of the form (5.35) can also be obtained for nuclei of nonmagnetic atoms for which HFI is described by expression (1.2a), but then A^2 will have to be replaced by $\sum_j A_j^2$, where the summation is carried over the nearest magnetic neighbors of the nonmagnetic atom under consideration.

Expression (5.35) illustrates the effect of exchange narrowing of a resonance line. Although the absolute magnitude $|\delta H|$ of the fluctuating field at the nucleus is $|AS/\gamma_n\hbar|$, the "lifetime" of this fluctuation τ_C (the characteristic correlation time) is very small because of the exchange interaction between the electron spins; according to (5.34) the correlation time is $\tau_C \sim \omega_E^{-1}$. As a result, the effectiveness of the fluctuating field $|\delta H|$ decreases roughly in proportion to the ratio of the corresponding Larmor precession period of the nucleus $t_L \sim (\gamma_n|\delta H|)^{-1} \sim (AS/\hbar)^{-1}$ to the correlation time $\tau_C \sim \omega_E^{-1}$. In other words, we have for the NMR line width

$$\Gamma_n = \frac{1}{T_2} \sim (AS/\hbar)\frac{\tau_C}{t_L} \sim (AS/\hbar)^2/\omega_E,$$

which coincides with the approximate part of equation (5.35).

The Neighborhood of the Curie Point

As is known from the theory of critical neutron scattering (Mori and Kawasaki, 1962), the time correlation of fluctuations in some types of electron spin motions increases strongly with the decrease of the temperature as the Curie (Néel) point is approached. This suppresses the effect of exchange narrowing and, consequently, leads to a broadening of the NMR line.

Using the method of Mori and Kawasaki (1962) for the investigation of the time correlation of spins in ferro- and antiferromagnets, Moriya (1962) developed a theory of nuclear magnetic relaxation in these materials for the temperature region near the magnetic transition point.

According to Moriya (1962), the asymptotic behavior of the line width and the rate of spin-lattice relaxation as $T \to T_C (T_N)$ is described by expressions of the form

$$\frac{1}{T_2} = \frac{1}{T_1} = \frac{C}{T_{1\infty}}\left(\frac{T_C}{T - T_C}\right)^{3/2} \quad \text{(ferromagnets),} \qquad (5.36)$$

$$\frac{1}{T_2} \approx \frac{1}{T_1} = \frac{C}{T_{1\infty}} \left(\frac{T_N}{T - T_N} \right)^{1/2} \quad \text{(antiferromagnets),} \tag{5.37}$$

where $T_{1\infty}^{-1}$ is the rate of longitudinal relaxation for $T \gg T_{C,N}$ determined by expression (5.35), while $C \sim 0.1$ is a numerical constant, which depends on the form of the lattice.*

As we approach the point of magnetic transition, a marked increase in the radius of the spatial correlation of electron spin fluctuations takes place. This in its turn leads to an increase in the indirect interaction between nuclear spins via electron spins (the analog of the Suhl–Nakamura interaction at high temperatures). However, in cubic crystals at high temperatures $T \gtrsim T_{C,N}$ (when the spontaneous magnetization is zero or small), this interaction will be isotropic, by analogy with the Ruderman–Kittel interaction (1.12). Therefore, if the interaction takes place between identical nuclei, it cannot contribute to the second moment of the NMR line (nor to its width). On the other hand, for identical nuclei in crystals with symmetry lower than cubic and for different nuclei irrespective of the symmetry type, the increase in the Suhl–Nakamura interaction as $T \to T_{C,N}$ can provide another reason for the appearance of singularities in the temperature dependence of the line width near the transition point. The theory of Moriya (1962) indicates that this contribution to the line widths for ferromagnets and antiferromagnets, on both sides of the transition point, is proportional to the quantity

$$\left(\frac{T_{C,N}}{|T - T_{C,N}|} \right)^{1/4} .$$

The experimental and theoretical investigations of the temperature dependence of NMR relaxation processes near the magnetic transition point offer another possibility for the study of critical phenomena in magnetic materials. Singularities in the temperature dependence of the line width, characteristic of phase transitions of second kind, have already been observed in a number of magnetic crystals.** Qualitatively they agree with the predictions of Moriya's theory (1962); a more detailed description of relaxation processes near transition points requires, however, further development of this theory. In particular, effects caused by the existence of near magnetic order must be taken into account.

---

* These expressions, strictly speaking, are applicable only to cubic crystals in the absence of an external field. Taking anisotropy and an external field into account leads to some complications, but the character of the singularity near $T_C(T_N)$ remains the same as before. Also note that the validity of these expressions is limited to the temperature intervals $1 \gg \dfrac{T - T_C}{T_C} > \left(\dfrac{\omega_e}{\omega_E} \right)^{1/2}$ or $1 \gg \dfrac{T - T_N}{T_N} > \dfrac{\omega_e}{\omega_E}$ for ferro- and antiferromagnets, respectively.

** See Review (1965a) (MnF_2), Heller and Benedek (1965a and b) (MnF_2 and EuS), Savatsky and Bloom (1964) and Abkowitz and Lowe (1966) ($CoCl_2 \cdot 6H_2O$).

Low Temperatures

The region of low temperatures, where the spin wave approximation is applicable, will be considered in more detail. For free spin waves (i.e., not interacting between themselves or with the lattice) we have, according to (2.12a),

$$\delta S_j^z = \pm (\langle a_j^+ a_j \rangle - a_j^+ a_j), \tag{5.38}$$

$$\delta S_j^\pm = (2S)^{1/2} a_j, \quad \delta S_j^\mp = (2S)^{1/2} a_j^+ . \tag{5.39}$$

In these expressions the upper sign corresponds to a ferromagnet; for antiferromagnets, the upper sign refers to one sublattice and the lower sign to the other. Since in all the expressions of this chapter, the main axis Z is the quantization axis for the nuclear spin, writing δS in this way, one implies that the quantization axes for the electron and the nuclear spins coincide.

Changing over from the operators a_j^+ and a_j to their Fourier components (and, if necessary, applying other unitary transformations which reduce the spin Hamiltonian to a sum of spin wave energies), we readily see that $\delta S_j^z(t)$ and, consequently, $f_0(t)$ contain components of the form $\exp[i(\omega_k - \omega_{k'})t]$. At the same time $\delta S_j^\pm(t)$ and, consequently, $f_1(t)$ are superpositions of harmonics of the form $\exp(\pm i\omega_k t)$ with the spin wave frequencies $\omega_k = \varepsilon_k/\hbar$.* Thus, the spectrum of longitudinal spin fluctuations δS_j^z contains arbitrarily low frequencies (as $\omega_k \to \omega_{k'}$), while the lowest spin wave (magnon) frequency for the transverse components $\delta S_j^\pm$ appears for $k = 0$ (i.e., the frequency of uniform ferro- or antiferromagnetic resonance, ω_e).

Since usually $\omega_e > \omega_n$, the real processes of emission or absorption of one magnon by a nucleus (which, as can be shown, are associated with the imaginary part of the function $_+\langle \delta S^+ | \delta S^- \rangle_{\omega_n}^{(S)}$) are forbidden by the energy conservation law. Accordingly, it follows from (5.12) that $1/T_1 = 0$; the real part of the function $_+\langle \delta S^+ | \delta S^- \rangle_{\omega_n}^{(S)}$ which describes virtual processes (emission of one magnon followed by its absorption by the same nucleus), will be nonzero, giving a shift in the NMR frequency. The pure imaginary Green's function $_+\langle \delta S^z | \delta S^z \rangle_{\omega=0}^{(S)}$ corresponds to real processes of magnon scattering at a nucleus without a change of energy, $\omega_{k'} = \omega_k$ (the so-called Raman scattering process), which also causes line broadening. As a result, in the free spin wave approximation we have $1/T_1 = 0$, $1/T_2$ is determined by the second

---

* Indeed, the Hamiltonian of the spin waves has the form $\mathscr{H}_s = \sum_{pk} \hbar\omega_{pk} a_{pk}^+ a_{pk}$, where a_{pk}^+ and a_{pk} are the creation and destruction operators for magnons of species p. The operators $\delta S^\pm$ are linear combinations of a_{pk}^+ and a_{pk}, so that their Heisenberg representation $\delta S^\pm(t) = \exp(i\mathscr{H}_s t/\hbar) \delta S^\pm \exp(-i\mathscr{H}_s t/\hbar)$ will contain only harmonics of the form $\exp(\pm i\omega_k t)$ since $a_k^+(t) = a_k^+ \exp(i\omega_k t)$ and $a_k(t) = a_k \exp(-i\omega_k t)$. At the same time δS_j^z is a quadratic form in a_{pk}^+ and a_{pk} and therefore the lowest harmonics entering δS_j^z will have the form $\exp[i(\omega_k - \omega_k)t]$.

term in (5.13) and, finally,

$$\delta\omega_n = (A/2\gamma_n\hbar)^2 \, \mathrm{Re} \, (_+\langle \delta S^+ \,|\, \delta S^- \rangle_{\omega_n}^{(S)}). \tag{5.30a}$$

As an example, consider the calculation of $\delta\omega_n$ and $1/T_2$ using these expressions for a ferromagnet. We obtain from (5.13) and (5.30a), taking (5.38) and (5.39) into consideration,

$$\delta\omega_n = \frac{A^2 S}{2\hbar^2 N} \sum_k \frac{1 + 2\bar{n}_k}{\omega_n - \omega_k}, \tag{5.40}$$

$$\frac{1}{T_2} = \left(\frac{A}{\hbar}\right)^2 \frac{\pi}{N^2} \sum_{k_1 k_2} \bar{n}_{k_1}(\bar{n}_{k_2} + 1)\, \delta(\omega_{k_1} - \omega_{k_2}). \tag{5.41}$$

Remember that $\bar{n}_k$ is the magnon Bose distribution function. The magnon frequency for small k, which constitutes the basic contribution to the sum, may again be written in the form

$$\omega_k = \omega_e + \omega_E(ak)^2. \tag{5.42}$$

If we pass from summation to integration and integrate taking (2.17) and (5.42) into account, we find (Kurkin and Turov, 1967)

$$\delta\omega_n = - \frac{SA^2\left(\sqrt{\omega_e} - \sqrt{\omega_e - \omega_n}\right)\kappa T}{4\pi\hbar^3 \omega_E^{3/2}\omega_n} \qquad (\omega_n < \omega_e), \tag{5.43}$$

$$\delta\omega_n \simeq - \frac{SA^2\kappa T}{8\pi\omega_E^{3/2}\,\omega_e^{1/2}\hbar^3} \qquad (\omega_n \ll \omega_e), \tag{5.43a}$$

$$\frac{1}{T_2} = \frac{A^2}{16\pi^3\hbar^2\omega_E}\left(\frac{\kappa T}{\hbar\omega_E}\right)^2 \ln\frac{\kappa T}{\hbar\omega_e}. \tag{5.44}$$

Here we have assumed $\kappa T \gg \hbar\omega_e$.[*]

Since the NMR frequency shift (5.43) depends on the temperature, it should be compared with the change $\delta\omega_n^{(\sigma)} = A\Delta\sigma(T)/\hbar$ caused by the temperature dependence of the mean spin $\sigma(T)$. Seeing that, in the first approximation,

$$\Delta\sigma(T) \simeq - 0.06\left(\frac{\kappa T}{\hbar\omega_E}\right)^{3/2},$$

we obtain

$$\frac{\delta\omega_n}{\delta\omega_n^{(\sigma)}} \sim \frac{2SA}{3\sqrt{\omega_e\kappa T}}. \tag{5.45}$$

---

[*] Expression (5.44) can be written also in the general case; it is only necessary to replace the factor $\ln(\kappa T/\hbar\omega_e)$ by $\ln(1 - e^{-(\hbar\omega_e/\kappa T)})^{-1}$. This expression was first derived, apparently, by Robert and Winter (1961).

Thus, if $\omega_n \ll \omega_e$, the corrections due to $\delta\omega_n$ (5.43) in the NMR determination of $\sigma(T)$ can be neglected. As a result of the dipole–dipole interaction, $\omega_e \sim 10^{10}$ sec^{-1} even in crystals with a low anisotropy. Since in the majority of cases $\omega_n \lesssim 10^9$ sec^{-1}, the above condition is usually fulfilled.

However, in some rare-earth elements, ω_n can be larger by several orders of magnitude (see Appendix II), which means that for the nuclei of these elements we may have $\omega_n \sim \omega_e$. In the latter case the NMR frequency shift $\delta\omega_n$ (5.43) may constitute a few tenths of $\delta\omega_n^{(\sigma)}$ (at $T \sim 10\hbar\omega_e/\kappa \sim 1°$K) and, consequently, it must be taken into account in the NMR determination of $\sigma(T)$.*

Expression (5.44) for the line half-width $\Gamma_n = 1/T_2$ gives $\Gamma_n/\omega_n \sim 10^{-8} T^2$ (degree)$^{-2}$ for $\omega_E = 10^{13}$ and $\omega_n = 10^9$ sec^{-1}. This mechanism of line broadening may play some role in the case of ferromagnetic dielectrics at moderately low temperatures. For example, an increase in $1/T_2$ proportional to T^2 was observed above $20°$K for the Fe^{57} nucleus in yttrium ferrite garnet (Robert and Winter, 1961).

A similar calculation of $\delta\omega_n$ and $1/T_2$ can be carried out for antiferromagnets (Moriya, 1956). For example, we have

$$\frac{1}{T_2} \simeq \frac{C}{6\pi\hbar^2} \frac{A^2}{\omega_E} \left(\frac{\kappa T}{\hbar\omega_E}\right)^3 \left[1 - I_1\left(\frac{\hbar\omega_e}{\kappa T}\right)\right], \qquad (5.46)$$

where C is a constant of the order of unity, determined by the number of spins in the unit cell and by its geometry;

$$I_1(x) = \frac{6}{\pi^2} \int_0^x \frac{y\,dy}{e^y - 1}.$$

We have again used an approximate expression for the frequency of the spin waves, corresponding to small wave vectors: $\omega_k^2 \simeq \omega_e^2 + \omega_E^2(ak)^2$. When the temperature is lowered, $1/T_2$ vanishes very rapidly, especially when $\hbar\omega_e > \kappa T$ (since $I_1(\infty) = 1$). In the case of an antiferromagnet, $1/T_2 \sim T^3$ for $\hbar\omega_e \ll \kappa T < \hbar\omega_E$.

In addition to one-magnon and two-magnon scattering processes, processes with the participation of three or more magnons should also be considered (Beeman and Pincus, 1967). In the temperature range $\kappa T \gg \hbar\omega_e$, for example, the processes of three-magnon scattering give $1/T_1 \sim T^{7/2}$ for ferromagnets and $1/T_1 \sim T^5$ for

---

* The frequency shift $\delta\omega_n$ may prove even more important in the investigation of deviations from the $T^{3/2}$ law. For example, the correction to the temperature dependence of $\sigma(T)$, caused by the presence of a gap (see Chapter II, Sec. 2), has the form

$$\Delta\sigma' = 0.08 \left(\frac{\kappa T}{\hbar\omega_E}\right)^{3/2} \left(\frac{\hbar\omega_e}{\kappa T}\right)^{1/2}$$

In this case

$$\frac{\delta\omega_n}{\delta\omega_n^{(\sigma)}} \sim \frac{SA}{2\hbar\omega_e}.$$

antiferromagnets. Another approach to the treatment of many-magnon scattering processes is described below in connection with the calculation of spin wave damping.

So far we have assumed that the quantization axes for electron and nuclear spins coincide. As a matter of fact, the angle θ_{Mm} between the equilibrium directions of the electron and the nuclear magnetization may differ from 0 or π although possibly by a small quantity; this deflection is associated with the fact that the total internal fields acting on the electron and the nuclear spins, respectively, are in general directed at some angle different from 0 or π to each other. For example, in the case of a uniaxial ferromagnet magnetized in the hard direction

$$| \sin \theta_{Mm} | \simeq \frac{H_a}{|H_n|} \frac{H}{H_a} \sqrt{1 - \left(\frac{H}{H_a}\right)^2} < \frac{H_a}{|H_n|} \tag{5.47}$$

(for $H \leq H_a$; in the case $H > H_a$, $\sin \theta_{Mm} = 0$). For an antiferromagnet magnetized perpendicularly to the antiferromagnetic axis

$$| \sin \theta_{Mm} | \simeq \left|\frac{H}{H_n}\right| . \tag{5.48}$$

If we take these circumstances into account, the processes of spin wave scattering by nuclei can contribute to the longitudinal relaxation, so that

$$\frac{1}{T_1} \simeq \frac{1}{T_2} \sin^2 \theta_{Mm} . \tag{5.49}$$

The latter follows from the fact that a transformation to components of S in the coordinate system where the Z axis is directed along the equilibrium electron magnetization gives rise to an admixture of the correlation function $f_0(t)$, proportional to $\sin^2 \theta_{Mm}$, in the expression (5.14) of the correlation function $f_1(t)$ for the transverse components S^+ and S^- (given in the coordinate system in which the Z axis is taken as the equilibrium direction of the nuclear magnetization).*

For nonmagnetic nuclei, the dipole fields of the surrounding magnetic atoms can contribute significantly to the local field (see expression (1.3)). It is therefore of interest to consider the relaxation caused by fluctuations in these dipole fields.

The dipole field fluctuations δH_j^D associated with spin waves are best expressed in this case in terms of a continuous magnetic medium, as a field set up by a magnetic space charge with a density $\rho_M = - \operatorname{div} M(r)$, namely

$$\delta H_j^D = - \nabla_j \int \frac{- \operatorname{div} M(r') \, dr'}{|R_j - r'|} , \tag{5.50}$$

where $M(r)$ is the local magnetization.

---

* The transformation to the electron coordinate system is necessary because it is in this coordinate system that relations (5.38) and (5.39) are valid.

The correlation functions entering (5.12) and (5.13) can easily be calculated using the continuum version of the spin wave theory (see, for example, Akhiezer et al., 1967). According to this theory, relation (2.12), which defines the transformation to the spin wave creation and destruction operators, takes for a ferromagnet the form

$$M_z(r) = M_0 - |\gamma_e|\hbar\frac{1}{V}\sum_{k_1 k_2} a_{k_1}^+ a_{k_2} e^{-i(k_1 - k_2)r}, \tag{5.51}$$

$$M^-(r) = \left(\frac{2|\gamma_e|\hbar M_0}{V}\right)^{1/2} \sum_k a_k e^{ikr},$$

$$M^+(r) = \left(\frac{2|\gamma_e|\hbar M_0}{V}\right)^{1/2} \sum_k a_k^+ e^{ikr} \tag{5.52}$$

(V is the volume of the specimen).

Remembering that one-magnon processes (emission and absorption of one spin wave by the nucleus) do not contribute to the relaxation rates (5.12) and (5.13), it is not difficult to see that only the term $\partial M_z/\partial z$ of div M must be retained in expression (5.50). Nevertheless, the fluctuating dipole field δH_j^D will contain both a longitudinal and a transverse component, determined by the fluctuations δM_z. We thus obtain from (5.12) and (5.13)

$$\frac{1}{T_2} = \frac{2}{T_1} = \frac{4(\gamma_n H_{\text{dip}})^2}{15\pi\omega_E}\left(\frac{\kappa T}{\hbar\omega_E}\right)^2 \ln\frac{\kappa T}{\hbar\omega_e}. \tag{5.53}$$

Here $H_{\text{dip}} \simeq \gamma_e\hbar/a^3$. Thus we should have $T_1 = 2T_2$ for local fields of the dipole type (while the temperature dependence remains the same as for the hyperfine relaxation mechanism). For antiferromagnets, the dipole mechanism gives in the same way $T_1 = 2T_2$, where T_2 is expressed by (5.36) with a quantity of the order of $\hbar\gamma_n H_{\text{dip}}$ inserted for the constant C (Moriya, 1956; Kranendonk and Bloom, 1956).

The Role of Spin Wave Damping

We have already mentioned that the energy conservation law forbids processes of nuclear magnetic relaxation through the emission or absorption of one magnon by the nucleus (for $\omega_e > \omega_n$). However, this exclusion refers, strictly speaking, only to an ideal magnon gas. The damping of magnons (spin waves) due to their interaction with one another, with the lattice, with impurities, etc., and the finite magnon lifetime introduce some uncertainty in the magnon energy and the above processes become possible.

Let us again consider a ferromagnet in which δH is produced by HFI and has the form (5.31). Taking spin wave damping into account, we obtain from (5.12) and (5.14) the following expression for $1/T_1$ (Kurkin and Turov, 1967):

$$\frac{1}{T_1} = \left(\frac{A}{\hbar}\right)^2 \frac{S}{N} \sum_k \frac{(1 + 2\bar{n}_k)\Gamma_k(\omega_n)}{(\omega_k - \omega_n)^2 + \Gamma_k^2(\omega_n)}. \tag{5.54}$$

Here $\Gamma_k(\omega_n)$ is a quantity which characterizes the damping of spin waves with a given wave vector k, at a frequency $\omega = \omega_n$. It is important to stress that $\Gamma_k(\omega_n)$ is a characteristic of spin wave damping at the NMR frequency, and not at the eigenfrequency $\omega = \omega_k$. Thus, we cannot take Γ_k directly from experiments on ferromagnetic resonance, spin wave resonance or other data in which the real spin wave damping at the eigenfrequency $\omega = \omega_k$ is determined. This quantity must be calculated theoretically.

The damping of spin waves may be caused by various factors. Kurkin and Turov (1967) considered the magnon–magnon interaction associated with the exchange energy of the electron spins and their magnetic anisotropy energy. It turned out that these damping mechanisms cannot contribute significantly to $1/T_1$ and $1/T_2$. Their contribution to the line width is small in comparison with expression (5.44), associated with Raman scattering of spin waves.

In some cases, the damping of spin waves caused by their scattering on impurities and other lattice defects can, apparently, have an appreciable value (De Gennes and Hartman-Boutron, 1961). Let $\omega_k \gg \Gamma_k, \omega_n$; then (5.54) takes the simpler form

$$\frac{1}{T_1} = \left(\frac{A}{\hbar}\right)^2 \frac{S}{N} \sum_k \frac{(1 + 2\bar{n}_k)\Gamma_k}{\omega_k^2}. \tag{5.55}$$

If we now assume that for small k, which give the principal contribution in the sum, Γ_k depends weakly on k, so that $\Gamma_k = \Gamma = $ const, we can replace the sum by an integral and, integrating (for $\omega_k = \omega_e + \omega_E(ak)^2$ and $\omega_e/\kappa T \ll 1$), we obtain

$$\frac{1}{T_1} \simeq \frac{1}{16\pi S} \frac{\omega_n^2}{(\omega_e\omega_E)^{1/2}} \left(\frac{\kappa T}{\hbar\omega_E}\right) \frac{\Gamma}{\omega_e}. \tag{5.56}$$

As is well known, impurities (rare earth ions) can play an important role in the damping of spin waves in yttrium ferrite garnet $Y_3Fe_5O_{12}$. In particular, this damping can be a determining factor in the broadening of the ferromagnetic resonance line at low temperatures. If we assume that it is the ratio Γ/ω_e which determines the relative half-width of the ferroelectric resonance line, there must be a direct proportionality between the rate of nuclear spin-lattice relaxation $1/T_1$ and this ratio according to (5.56). And indeed, such a correlation was observed between $1/T_1$ and Γ in yttrium ferrite garnet at low temperatures (where rare earth impurities play an important role in the NMR line broadening). The width of the NMR line contributed by the rare earth impurities has a characteristic maximum in its temperature dependence, and this maximum is reproduced, roughly at the same temperature, in the temperature dependence of the spin-lattice relaxation rate $1/T_1$ of the Fe^{57} nuclei (Robert and Winter, 1961).

Experiment, in agreement with theory, shows that, in the first approximation, Γ/ω_e does not depend on the field H, and, consequently, the field dependence of the spin-lattice relaxation time squared is given, according to (5.56), by the relation

$$T_1^2 \sim \omega_e = |\gamma_e|\left(H + H_A - \frac{4\pi}{3}M_S\right),\tag{5.57}$$

i.e., it is linear.* This is also corroborated by experiment.

An analogous relationship is found between the longitudinal relaxation of the nuclei $1/T_1$ and the transverse relaxation of the electrons $1/T_{2e} = \Gamma$ in manganese ferrite spinel $MnFe_2O_4$ (Blocker and Heeger, 1966). Here the iron ions with different valencies play the role of impurities (admixture of Fe^{2+} ions to Fe^{3+} ions). However, our treatment based on expression (5.54) has a purely qualitative character and does not take into consideration the specific features of these cases. A more detailed discussion on the role of impurities in nuclear spin-lattice relaxation is given in the previously cited references.

Mizoguchi and Inoue (1966) showed that the presence of ions with different valencies and the resulting likelihood of electron migration significantly affect the nuclear magnetic relaxation in magnetite Fe_3O_4.

6. Fluctuations of the Local Fields Created by Conduction Electrons

In metallic ferromagnets the nuclear spins can interact not only with the collective spin vibrations of the electron system—the spin waves—but also with individual Fermi excitations, viz., the collective electrons or the conduction electrons. Accordingly, for metals of the iron group this involves both the 4s-band electrons, which are basically responsible for electrical conductivity, and the 3d-band electrons, which are the cause of ferromagnetism and are also mobile.

Let us introduce second-quantized electron wave functions

$$\psi(r, s) = \sum_v a_v \psi_v(r, s),$$

$$\psi^+(r, s) = \sum_v a_v^+ \psi_v^*(r, s),\tag{5.58}$$

where a_v and a_v^+ are respectively the electron destruction and creation operators in the state $v \equiv nk\sigma$ (n is the number of the band, k and σ are the electron quasi-momentum and spin), while $\psi_v(r, s)$ is the wave function of this state (r and s are the spatial and spin coordinates of the electron). In the absence of spin-orbit coupling,

---

* Expression (5.57) evidently takes into account the demagnetizing field $-(4\pi/3)M_S$ for a spherical specimen, which enters the frequency ω_e defined as the limit of ω_k as $k \to 0$.

$\psi_v(r, s)$ can be represented in the form of the product of a spin and a coordinate part

$$\psi_v(r, s) = C_\sigma(s)\, \psi_{nk}(r), \tag{5.59}$$

where $C_\sigma(s)$ is the spin wave function, while $\psi_{nk}(r)$ is the Bloch wave function for the band n.

Let us first consider the longitudinal relaxation. It is caused by the fluctuating fields at the nucleus, which constitute the mean (quantum-mechanical) values of the transverse components of the local electron field operator H_e (1.1) acting on the nuclear spin:

$$\delta H = \sum_s \int \psi^+(r, s)\, H_e \psi(r, s)\, dr,$$

or taking (5.58) into account

$$\delta H^+ = \sum_{vv'} \alpha(v, v')\, a_v^+ a_{v'}, \qquad \delta H^- = \sum_{vv'} \alpha^*(v, v')\, a_v^+ a_{v'}, \tag{5.60}$$

where

$$\alpha(v, v') = \sum_s \int \psi_v^*(r, s)\, H_e^+ \psi_{v'}(r, s)\, dr. \tag{5.61}$$

It is a characteristic property of the scattering processes of conduction electrons by nuclear spins that, irrespective of the interaction mechanism, they lead to a linear temperature dependence of the longitudinal relaxation rate. Indeed, we have according to (5.12), (5.14) and (5.60)

$$\frac{1}{T_1} = \pi\hbar\gamma_n^2 \sum_{vv'} |\alpha(v, v')|^2 f_v(1 - f_v)\, \delta(E_v - E_{v'}), \tag{5.62}$$

where E_v is the electron energy in the state v, while f_v is the Fermi distribution function

$$f_v \equiv \langle a_v^+ a_v \rangle \equiv [\exp[(E_v - \zeta)/\kappa T] + 1]^{-1}$$

(ζ is the chemical potential). Since

$$f_v(1 - f_v) = -\kappa T \frac{df_v}{dE_v} \simeq \kappa T \delta(E - E_F),$$

we finally obtain instead of (5.62)

$$\frac{1}{T_1} = \pi\hbar\gamma_n^2 \kappa T \sum_{n\sigma,\, n'\sigma'} \langle |\alpha(nk\sigma, n'k'\sigma')|^2 \rangle_F g_{n\sigma}(E_F)\, g_{n'\sigma'}(E_F), \tag{5.63}$$

where $E_F \simeq \zeta$ is the Fermi energy, $g_{n\sigma}(E)$ is the density of states for electrons of

species $n\sigma$, and $\langle \ldots \rangle_F$ is the mean value of the squared modulus of the matrix element $\alpha(v, v')$ on the surfaces $E_{nk\sigma} = E_F$ and $E_{n'k'\sigma'} = E_F$.*

Using the explicit form (1.1) of the local field operator at the nucleus, the relaxation rate $1/T_1$ associated with one group of conduction electrons (for example, with the electrons of the s- or d-band in ferromagnets of the iron group) can be found. This generally requires knowledge of the wave functions and the energy spectrum of these electrons. An estimate obtained by Moriya (1964) showed that in the case of iron, cobalt and nickel, the largest contribution to $1/T_1$ came from the orbital field set up at the nucleus by d-band electrons (the first term with l in expression (1.1) for H_e). In the strong coupling approximation, Moriya (1964) obtained for the d-electrons from (5.63), taking (5.59) and (5.61) into account,

$$\left(\frac{1}{T_1}\right)_d \simeq \frac{16}{5}\, \pi\hbar\kappa T \gamma_n^2 \mu_B^2 \sum_\sigma [\langle 1/r^3 \rangle_{F\sigma} g_{d\sigma}(E_F)]^2 , \qquad (5.64)$$

where $\langle 1/r^3 \rangle_{F\sigma}$ is a matrix element averaged over the Fermi surface for d-band electrons of spin σ.

A quantitative estimate based on expression (5.64) (taking into account the correction of Kaplan et al. (1966)) leads to the relation

$$(\gamma_n^2 T_1 T)_d^{-1} \sim 1\cdot 10^7 \qquad (5.65)$$

(with variations by a factor of a few times in either direction for various ferromagnets). The linear temperature dependence of $1/T_1$ is indeed observed in Fe, Co and Ni (Weger et al., 1961; Volsted and Wernick, 1966).

At the same time, the experimental value of $1/T_1$ obtained in the work of Weger et al. (1961) is by a factor of 2–10 greater than the theoretical estimates of Moriya (1964) and Kaplan et al. (1966). However, these experimental values refer to the rate of longitudinal relaxation of nuclei in domains in the absence of an external magnetizing field. It can be assumed that the presence of domains and domain walls in these conditions may significantly influence the spin-lattice relaxation of the nuclei (see Chapter VII).

In the work of Jaccarino et al. (1966), the field dependence of $1/T_1$ in ferromagnetic metals was investigated experimentally. The magnitude of $1/T_1$ was found to decrease with increasing H (remaining a linear function of T) and reached a constant value at $H > 4\pi M_S$ (Figure 12). It is this quantity, being the true characteristic of spin-lattice relaxation in single-domain specimens, that should be compared with the theoretical predictions given by relation (5.64). And indeed, the experimental values of

---

* By definition

$$\langle |\alpha(nk\sigma, n'k'\sigma')|^2 \rangle_F = g_{n\sigma}^{-1}(E_F)\, g_{n'\sigma'}^{-1}(E_F) \int |\alpha(nk\sigma, n'k'\sigma')|^2\, g_{n\sigma}(E_F, \mathscr{F})\, g_{n'\sigma'}(E_F, \mathscr{F}')\, d\mathscr{F}\, d\mathscr{F}' ,$$

where $g_{n\sigma}(E, \mathscr{F})\, d\mathscr{F}$ is the number of states in k-space falling in a unit energy interval on a surface element $d\mathscr{F}$ around a point $\mathscr{F}$ on the surface of equal energy $E_{nk\sigma} = E$. Note that $\int g_{n\sigma}(E, \mathscr{F})\, d\mathscr{F} = g_{n\sigma}(E)$.

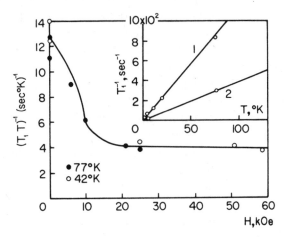

Figure 12

The dependence of $(T_1 T)^{-1}$ on the magnetic field for the Co^{59} nuclei in metallic cobalt.
The inset shows the temperature dependence of $1/T_1$ for the fields $H = 0$ (curve 1) and
$H = 25$ kOe (curve 2).

$1/T_1$ at high fields coincide, in order of magnitude, with the theoretical estimates
(5.65).

For metals of the iron group, the two other terms of the local field (1.1), representing the dipole interaction between the nuclear spin and the d-band electrons
and the Fermi contact interaction with the s-band electrons, respectively, contribute, according to the estimate of Moriya (1964), considerably less than the orbital
field.

However, according to Matsuura (1966), the Fermi contact interaction with the
$4s$ conduction electrons is the main mechanism of $1/T_1$ in the ferromagnet Mn_4N,
in the temperature interval between $1.8°$ and $77°K$.

Weger et al. (1961) and Moriya (1964) also mentioned two indirect mechanisms
for the interaction of nuclear spins with conduction electrons. These are the interaction with the d-band electrons via electrons of the internal shells, as a result of the
exchange polarization of the latter, and the interaction with the s-band electrons
via spin waves. The longitudinal relaxation rate $1/T_1$ due to both these mechanisms,
is again a linear function of temperature. However, their contribution to the magnitude of $1/T_1$ is apparently less than the contribution of the direct orbital field of the
$3d$-electrons.

Finally, let us briefly consider the problem of the transverse relaxation rate $1/T_2$
produced by the fluctuations of the local field of the conduction electrons (Kaplan
et al., 1966).

In the absence of spin-orbit interaction, the correlation functions of the d-electron

orbital fields must be isotropic;

$$\langle \delta H^x(t)\, \delta H^x(0)\rangle = \langle \delta H^y(t)\, \delta H^y(0)\rangle = \langle \delta H^z(t)\, \delta H^z(0)\rangle \,.\,*$$

It is not difficult to see from expressions (5.12)–(5.15) that in this case $(1/T_2)_d = (1/T_1)_d$. The linear temperature dependence of $1/T_2$, which follows from these relations, was observed experimentally for the Mn^{55} impurity nuclei in iron (Kaplan et al., 1966) and for the Ni^{61} nuclei in nickel (Weger et al., 1961). The longitudinal and the transverse relaxation rates were not equal: their ratio, in the two works mentioned, was roughly equal to 5 and 3, respectively. It must be borne in mind, however, that these experimental results refer again to a case in which no external field is applied, i.e., the case of domain structure, while the theoretical relation $T_1 = T_2$ is obtained for a single-domain specimen.

7. The Spin–Phonon Interaction

In addition to spin wave damping, another mechanism for one-magnon nuclear spin relaxation processes is the "mixing", due to the magneto-elastic interaction, of electron spin vibrations (magnons) with lattice vibrations (phonons). Actually, the eigenvibrations are coupled magnon–phonon waves. The spectrum of these waves is shown schematically in Figure 13. The lower ("quasi-magnon") branch has no energy gap and contains, in particular, quanta of NMR frequency which can be emitted and absorbed by nuclear spins.

A calculation of the rate of these direct processes of relaxation by "quasi-magnons" was carried out for antiferromagnets of the EA type by Pincus and Winter (1963). However, the treatment based on the above concepts requires very tedious computations of the spectrum of the coupled magnon–phonon waves. Actually, at the relevant frequencies ($\omega \sim \omega_n \ll \omega_e$) the coupling of these waves is usually very weak, so that it can be treated by pertubation theory, the spin waves being considered as an intermediary subsystem through which the nuclear spins interact with the phonons. This approach was developed by Silverstein (1963) and Buishvili (1963b) in their treatment of the nuclear magneto-acoustic resonance in ferro- and antiferromagnets (see Chapter IX). Because of its simplicity, this approach will be the basis for the discussion which follows. In addition, a number of other possible interaction mechanisms between nuclear spins and lattice vibrations will be discussed.

---

* Indeed, according to (1.1) and (5.58), we obtain for the component i of the orbital field

$$\delta H^i = - \gamma_e \hbar \sum_{nk\sigma,\, n'k'} a^+_{nk\sigma} a_{n'k'\sigma} \int \psi^+_{nk}(r)\, \frac{l_i}{r^3}\, \psi_{n'k'}(r)\, dr\,.$$

It is obvious that the matrix elements of all the three components of the orbital field operator l_i/r^3, which are represented by the integral in the last expression, are symmetric (at least for a cubic crystal), whence the isotropy of the correlation functions follows.

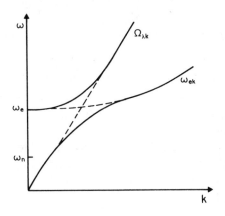

Figure 13

The spectrum of coupled magneto-acoustic waves.
$\Omega_{\lambda k}$ is the quasi-phonon branch, ω_{ek} is the quasi-magnon branch.

Interaction Mechanisms of Nuclear Spins with the Lattice

Let us consider a continuous magnetic elastic medium, characterized by local magnetizations $M_p(r)$ (the number of values of p is equal to the number of magnetic sublattices needed for the description of the magnetic properties of the medium) and elastic displacements $u(r)$ of its points. The potential energy density $\mathscr{F}$ of this medium is a function of the magnetizations M_p, their derivatives $\partial M_{p\alpha}/\partial x_\beta$, and the strain tensor components $\partial u_\alpha/\partial x_\beta$. However, these quantities do not enter $\mathscr{F}$ arbitrarily but only in the form of definite combinations, ensuring the invariance of $\mathscr{F}$ under arbitrary rotations of the magnetic crystal which rotate the lattice and the magnetic moments as one rigid system (Tiersten, 1964; Akhiezer et al., 1967).

It thus turns out that the magneto-elastic part of $\mathscr{F}$, in the linear approximation with respect to deformation, can be represented as a sum of terms of two types (Vlasov, 1960; Bar'yakhtar et al., 1964):

$$\mathscr{F} = \mathscr{F}^{(u)} + \mathscr{F}^{(\varepsilon)}. \tag{5.66}$$

Terms of the first type $\mathscr{F}^{(u)}$ contain only the symmetrical part of the distortion tensor, i.e., the strain tensor

$$u_{\alpha\beta} = \frac{1}{2}\left(\frac{\partial u_\alpha}{\partial x_\beta} + \frac{\partial u_\beta}{\partial x_\alpha}\right), \tag{5.67}$$

and represent the ordinary magnetostriction (and, in general, piezomagnetism)

$$\mathscr{F}^{(u)} = \lambda_{\alpha\beta}(M_p)\,u_{\alpha\beta}. \tag{5.68}$$

Here $\lambda_{\alpha\beta}$ is some tensorial function of the magnetizations M_p, the exact form of which

is determined by the symmetry of the crystal and the magnetic structure; α, $\beta \equiv$ x, y, z; summation over repeating indices is implied. Terms of the second type are fully determined by the form of the magnetic anisotropy energy $a(M_p)$. They are obtained as the increments of $a(M_p)$ due to small rotations of a volume element characterized by the antisymmetric part of the distortion tensor, i.e., the rotation tensor

$$\varepsilon_{\alpha\beta} = \frac{1}{2}\left(\frac{\partial u_\alpha}{\partial x_\beta} - \frac{\partial u_\beta}{\partial x_\alpha}\right). \tag{5.69}$$

When the coordinate system fixed to the relevant volume is turned through an infinitesimal angle, the vector components transform in the following manner:

$$M_{p\beta} \rightarrow M_{p\beta} + M_{p\alpha}\varepsilon_{\alpha\beta}. \tag{5.70}$$

The local change in the magnetic anisotropy energy caused by the rotations is given, in the linear approximation in $\varepsilon_{\alpha\beta}$, by the expression

$$\mathscr{F}^{(\varepsilon)} = a(M_{p\beta} + M_{p\alpha}\varepsilon_{\alpha\beta}) - a(M_{p\beta}) \simeq \frac{\partial a}{\partial M_{p\beta}} M_{p\alpha}\varepsilon_{\alpha\beta}. \tag{5.71}$$

Thus, due to the magneto-elastic energies (5.68) and (5.71), the electron spins interact with lattice vibrations described by the tensors $u_{\alpha\beta}$ and $\varepsilon_{\alpha\beta}$. On the other hand, the electron spins also interact with the nuclear spins, as a result of HFI. In terms of a continuous medium, the HFI energy density (1.2a) per one nuclear spin I_l can be presented in the form

$$\mathscr{H}_{\mathrm{HFI}}(r) = \gamma_n \sum_p A_p(r - R_l) I_l M_p(r). \tag{5.72}$$

The nuclear spins I_l will thus interact with the lattice vibrations $u_{\alpha\beta}$ and $\varepsilon_{\alpha\beta}$ via the magnetization M_p.

Using the explicit form of the magneto-striction energy and magnetic anisotropy energy for a particular magnetic crystal, we must subsequently isolate in (5.68) and (5.71) the linear terms with respect to the transverse magnetization components M_{px} and M_{py}, assuming also $M_{pz} \simeq M_{p0}$ (the Z axis coincides with the equilibrium direction of the p-th sublattice magnetization).

Let us first consider a simple cubic or uniaxial ferromagnet, magnetized along the easy axis, which coincides with an edge of the cube or with the main symmetry axis, respectively. It is easy to see that in both cases an expression of the following form is obtained from (5.68) and (5.71):

$$\mathscr{F}(r) \simeq (Gu_{zx} + 2K_1\varepsilon_{zx})\frac{M_x}{M_0} + (Gu_{zy} + 2K_1\varepsilon_{zy})\frac{M_y}{M_0}, \tag{5.73}$$

where G is one of the magnetostriction constants and K_1 is the first constant of magnetic anisotropy.*

Selecting linear terms with respect to M_x and M_y in the HFI energy (5.72), we pass, using (5.52), to the second quantization representation and express the total magneto-elastic energy $\int \mathscr{F}(r)\,dr$ and the HFI energy $\int \mathscr{H}_{HFI}(r)\,dr$ via the spin wave creation and destruction operators a_k^+ and a_k. It is then easy to find the effective Hamiltonian $\mathscr{H}_{IL}$ of the indirect interaction between nuclear spins and the lattice via spin waves. Such an interaction is effected by means of virtual processes of one-magnon emission by the lattice (representable by $u_{\alpha\beta}$ and $\varepsilon_{\alpha\beta}$) followed by absorption by a nucleus, or vice versa. The calculation can be carried out according to perturbation theory in the same manner as in Chapter IV for the indirect interaction between nuclear spins via spin waves, or by the method of canonic transformations with subsequent elimination of the variables of the "intermediary" electron system by an appropriate averaging (Silverstein, 1963). As a result we obtain for a ferromagnet

$$\mathscr{H}_{IL} = -\frac{1}{2}\gamma_n \hbar H_n \left[I_l^+ \sum_k \frac{\omega_{ms} u_{z-}(k) + \omega_A \varepsilon_{z-}(k)}{\omega_k} e^{-ikR_l} + \text{compl. conj.} \right], \quad (5.74)$$

where in the general case

$$H_n = -M_0 \int A(r)\,dr \equiv \sum_j A_j \langle S_j^z \rangle / \hbar\gamma_n,$$

while for the nuclei of magnetic atoms, $H_n = -A_0 M_0 \equiv -A\sigma/\hbar\gamma_n$;

$$\omega_{ms} = |\gamma_e|(G/M_0), \quad \omega_a = |\gamma_e| H_a = |\gamma_e|(2K_1/M_0), \quad (5.75)$$

$$u_{z\pm} = u_{zx} \pm i u_{zy},$$

$$u_{\alpha\beta}(k) = \frac{1}{V}\int u_{\alpha\beta}(r) e^{ikr}\,dr\,; \quad (5.76)$$

$\omega_k = \omega_e + \omega_E(ak)^2$ is the spin wave frequency.

An analogous, though much more complicated calculation can be carried out for an antiferromagnet of the EA type (Silverstein, 1963; Turov and Timofeev, 1967).

---

* For example, for a cubic crystal, we proceed from the expressions

$$\mathscr{F}^{(u)} = (G_0/M_0^2) u_{\alpha\alpha} M_\alpha^2 + (G/2M_0^2) u_{\alpha\beta} M_\alpha M_\beta \quad (\alpha \neq \beta),$$

$$a(M) = (K_1/M_0^4)(M_x^2 M_y^2 + M_x^2 M_z^2 + M_y^2 M_z^2).$$

However, for cubic crystals we usually have $K_1 \ll G$, so that the "rotation" mechanism of magneto-elastic interaction does not play a significant role. For uniaxial crystals or for crystals with lower symmetry, K_1 and G in general can be of the same order of magnitude.

In the case of a moderately large field H (such that $H^2 \ll H_E H_a$) directed along the antiferromagnetic axis $L \parallel Z$, we obtain in the same approximation

$$\mathscr{H}_{IL} = -\frac{1}{2}\gamma_n \hbar I_l^+ \left\{ H_n \sum_k \frac{\gamma_e H_E \left[\omega_{ms} u_{z-}(k) + \omega_a \varepsilon_{z-}(k) \right]}{\omega_{1k}\omega_{2k}} e^{-ikR_l} \right\} + \text{compl. conj.}$$

$$(5.77)$$

Here $\omega_{1k} \simeq \omega_{2k} \simeq \omega_k = \sqrt{\gamma_e^2 H_E H_a + \omega_E^2 (ak)^2}$ is the spin wave energy, $H_n =$

$- M_0 \int [A_1(r) - A_2(r)] \, dr$ is the total hyperfine field contributed by the two sub-

lattices and acting on the nuclear spin. In terms of a discrete medium,

$$H_n = \sum_j A_j \langle S_j^z \rangle / \hbar \gamma_n .$$

Finally, for an antiferromagnet of the EP type, if again the Y axis is directed along the antiferromagnetic axis and the X axis is parallel to the external magnetic field and to the resultant magnetic moment (see Figure 7), the effective Hamiltonian of the indirect interaction between nuclear spins and the lattice can be presented in the form

$$\mathscr{H}_{IL} = -\frac{1}{2}\gamma_n \hbar I_l^+ \left\{ - iH_n^\zeta \sum_k \frac{\omega_{ms}^{(1)} |\gamma_e| H_E}{\omega_{1k}^2} u_{xy}(k) e^{-ikR_l} + \right.$$

$$\left. + H_n^y \sum_k \frac{\left[\omega_{ms}^{(2)} u_{yz}(k) + \omega_a \varepsilon_{yz}(k) \right] |\gamma_e| H_E}{\omega_{2k}^2} e^{-ikR_l} \right\} + \text{compl. conj.} \qquad (5.78)$$

Here the frequencies

$$\omega_{(1,2)k} = \left[\omega_{e(1,2)}^2 + \omega_E^2 (ak)^2 \right]^{1/2} \qquad (5.79)$$

are the "low" and the "high" spin wave frequencies, respectively (see Chapter II); ω_{e1} and ω_{e2} are determined for the various cases by expressions (2.68)–(2.69), (2.83)–(2.84), or (2.86)–(2.87). H_n^ζ and H_n^y are the projections of the hyperfine field

$$H_n = - M_0 \int [\zeta_1 A_1(r) + \zeta_2 A_2(r)] \, dr \equiv \sum_j A_j \langle S_j \rangle / \gamma \hbar \qquad (5.80)$$

on the axis of nuclear spin quantization* and on the Y axis, respectively (ζ_1 and ζ_2 are unit vectors in the direction of the equilibrium magnetizations M_1 and M_2). The circular projections of the nuclear spin $I_l^\pm$ in expression (5.78) are written as $I_l^\pm = I_l^\xi \pm iI_l^\eta$, where ξ and η are the unit vectors of the axes perpendicular to ζ (ξ can without loss of generality be directed along Z). Expression (5.78) contains two

---

* In the general case when there are both a hyperfine field (5.80) and an external field H,

$$\zeta = (H + H_n)/|H + H_n|.$$

magnetostriction frequencies $\omega_{ms}^{(1)}$ and $\omega_{ms}^{(2)}$, which are determined from the magneto-striction constants G_1 and G_2 as in (5.75).*

We have so far assumed that the HFI is isotropic. In this case the indirect interaction between nuclear spins and lattice vibrations via spin waves is in fact the only interaction mechanism between these subsystems in dielectrics (if the quadrupole interaction is ignored, see Chapter VI). For anisotropic HFI of the form

$$\text{(summation over repeating indices)} \qquad (5.81)$$

(which includes also the dipole interaction as a special case), there can exist another spin-lattice coupling mechanism associated with the modulation of the tensor $A_{lj}^{\alpha\beta} \equiv A^{\alpha\beta}(\boldsymbol{R}_{lj})$ by lattice vibrations. Because $A_{lj}^{\alpha\beta}$ depends on the distance $\boldsymbol{R}_{lj} = \boldsymbol{R}_l - \boldsymbol{R}_j$ between the nucleus l and the surrounding magnetic atoms j, the change in the HFI energy produced by lattice vibrations can be written, in the linear approximation with respect to strain, in the form

$$\mathscr{H}_{IL} = \Delta\mathscr{H}_{HFI}^{(u)} = \frac{\partial A^{\alpha\beta}}{\partial R_{lj}^{\gamma}} R_{lj}^{\delta} u_{\gamma\delta} I_l^{\alpha} S_j^{\beta}. \qquad (5.82)$$

If some components of the HFI tensor do not vanish for $\alpha \neq \beta$, $\beta = z$ (z is the electron spin quantization axis), the hyperfine field at the nucleus will contain, for $S^z \sim S$, variable transverse components proportional to $u_{\gamma\delta}$. This leads to a direct interaction between the transverse nuclear spin component and the lattice vibrations. All this applies also to the dipole–dipole interaction between nuclear spins and the surrounding magnetic atoms.

In analogy to the magneto-elastic interaction, the anisotropic HFI varies not only as a result of deformation vibrations (characterized by the tensor $u_{\alpha\beta}$) but also owing to rotational vibrations (characterized by the tensor $\varepsilon_{\alpha\beta}$). If we remember that the vector components I_l^{α} and S_j^{β} in (5.81) transform like (5.70) when the coordinate system fixed to the volume element is rotated through a small angle, the change in the HFI energy caused by the rotation of the volume element which contains

---

* G_1 and G_2 are the constants before the magneto-elastic terms of the form $(G_1/M_0^2)\,L_xL_yu_{xy} + (G_2/2M_0^2)\cdot$ $(L_yu_{yz} + L_xu_{xz})\,L_z$. In general, for a rhombohedral crystal of the MnCO$_3$ (or α-Fe$_2$O$_3$) type, say, expression (5.78) should contain four magnetostriction constants (Turov and Shavrov, 1965). Besides these terms, which do not lead to a dependence of $\mathscr{H}_{IL}$ on the direction of $\boldsymbol{H}$ in the basal plane, the magneto-elastic energy also contains two terms (from among the linear terms in L_x and L_z) which depend on the direction of $\boldsymbol{H}$:

$$(G_3/M_0^2)\,L_xL_y(\cos 3\psi_Hu_{zx} - \sin 3\psi_Hu_{zy}) + (G_4/2M_0^2)\,L_xL_y\,[(u_{xx} - u_{yy})\cos 3\psi_H - 2u_{xy}\sin 3\psi_H],$$

where ψ_H is the azimuth angle determining the direction of $\boldsymbol{H}$ in the basal-plane (we recall that the equilibrium directions of $\boldsymbol{M} = \boldsymbol{M}_1 + \boldsymbol{M}_2$ and $\boldsymbol{L} = \boldsymbol{M}_1 - \boldsymbol{M}_2$ are along the X and Y axis, respectively). Assuming that the magnetostriction anisotropy in the basal plane is small, we are retaining in the magneto-elastic energy only terms with G_1 and G_2.

the nuclear spins I_l can be written, in a linear approximation in $\varepsilon_{\alpha\beta}$, as

$$\mathscr{H}_{IL} \equiv \Delta \mathscr{H}_{HFI}^{(\varepsilon)} = (A_{lj}^{\alpha\delta}\varepsilon_{\delta\beta} - A_{lj}^{\delta\beta}\varepsilon_{\alpha\delta})\, I_l^\alpha S_j^\beta . \tag{5.83}$$

In the special case when the HFI is purely local and the HFI tensor is diagonal, so that

$$A_{lj}^{\alpha\beta} = A_\alpha \delta_{lj}\delta_{\alpha\beta} , \tag{5.84}$$

we obtain instead of (5.83)

$$\mathscr{H}_{IL} = (A_\alpha - A_\beta)\,\varepsilon_{\alpha\beta} I^\alpha S^\beta . \tag{5.85}$$

One-Phonon Processes

Representing the Hamiltonian $\mathscr{H}_{IL}$ (more precisely, the terms with $\alpha = x,\, y$ and $\beta = z$) for any spin-lattice coupling mechanism in the form

$$\mathscr{H}_{IL} = -\tfrac{1}{2}\gamma_n \hbar (I_l^+ \delta H_l^- + I_l^- \delta H_l^+), \tag{5.86}$$

we can write down the fluctuating field $\delta H_l^\pm$ associated with lattice vibrations, and then calculate, from expressions (5.12) and (5.14), the corresponding longitudinal (spin-lattice) relaxation rate.

For example, for the mechanism of indirect interaction of nuclei with the lattice via spin waves, we have from (5.74), (5.77) and (5.78) for a ferromagnet (F), an antiferromagnet of the EA type (AF–EA) and an antiferromagnet of the EP type (AF–EP):

$$\text{(F)} \quad \delta H_l^\pm = H_n \sum_k e^{\pm i\mathbf{k}\mathbf{R}_l}\, \frac{\omega_{ms}u_{z\pm}(\mp\mathbf{k}) + \omega_a \varepsilon_{z\pm}(\mp\mathbf{k})}{\omega_k} , \tag{5.87}$$

$$\text{(AF–EA)} \quad \delta H_l^\pm = H_n \sum_k e^{\pm i\mathbf{k}\mathbf{R}_l}\, \frac{|\gamma_e|\, H_E\left[\omega_{ms}u_{z\pm}(\mp\mathbf{k}) + \omega_A \varepsilon_{z\pm}(\mp\mathbf{k})\right]}{\omega_k^2} , \tag{5.88}$$

$$\text{(AF–EP)} \quad \delta H_l^\pm = \pm i H_n^\zeta \sum_k e^{\pm i\mathbf{k}\mathbf{R}_l}\, \frac{\omega_{ms}^{(1)}|\gamma_e|\, H_E u_{xy}(\mp\mathbf{k})}{\omega_{1k}^2} +$$

$$+ H_n^y \sum_k e^{\pm i\mathbf{k}\mathbf{R}_l}\, \frac{|\gamma_e|\, H_E\left[\omega_{ms}^{(2)}u_{yz}(\mp\mathbf{k}) + \omega_A \varepsilon_{yz}(\mp\mathbf{k})\right]}{\omega_{2k}^2} . \tag{5.89}$$

Being linear in deformation (the tensor components $u_{\alpha\beta}$ and $\varepsilon_{\alpha\beta}$), these fields produce one-phonon spin-lattice relaxation processes, i.e., absorption and emission of one phonon by the nucleus. In the calculation of $1/T_1$ from (5.14) and (5.12), it is convenient to use again the second quantization representation. For this purpose we introduce the creation and destruction operators $b_{q\lambda}^+$ and $b_{q\lambda}$ in place of the displacement vector u (see, for example, Ziman (1966), Chapter II),

$$u(r) = \frac{1}{\sqrt{V}} \sum_{q\lambda} e_{q\lambda} \left(\frac{\hbar}{2\rho\Omega_{q\lambda}}\right)^{1/2} (b_{q\lambda} e^{-iqr} + b_{q\lambda}^+ e^{iqr}), \qquad (5.90)$$

where $\Omega_{q\lambda}$ is the frequency of a phonon with wave vector q and polarization λ ($\lambda = 1, 2, 3$), $e_{q\lambda}$ is the polarization unit vector and ρ is the density of the material. For the sake of simplicity we shall assume that $\Omega_{q\lambda} = c_s q$ for all three polarizations, where c_s is the velocity of sound.

The Hamiltonian of the phonon system has the form

$$\mathscr{H}_L = \sum_{q\lambda} (N_{q\lambda} + \tfrac{1}{2}) \hbar\Omega_{q\lambda}, \qquad (5.91)$$

where $N_{q\lambda} = b_{q\lambda}^+ b_{q\lambda}$ is the phonon number operator for the state $q\lambda$. In the Heisenberg representation (figuring in (5.14))

$$b_{q\lambda}(t) = e^{(i/\hbar)\mathscr{H}_L t} b_{q\lambda} e^{-(i/\hbar)\mathscr{H}_L t} = b_{q\lambda} e^{-i\Omega_q \lambda t},$$

$$b_{q\lambda}^+(t) = b_{q\lambda}^+ e^{i\Omega_q \lambda t}. \qquad (5.92)$$

The mean values are

$$\langle b_{q\lambda}^+ b_{q'\lambda'} \rangle = N_{q\lambda} \delta_{q'q} \delta_{\lambda'\lambda} \quad \text{and} \quad \langle b_{q\lambda} b_{q'\lambda'}^+ \rangle = (1 + \bar{N}_{q\lambda}) \delta_{q'q} \delta_{\lambda'\lambda}, \qquad (5.93)$$

where $\bar{N}_{q\lambda}$ is the Bose distribution function of the form (2.17) for particles of energy $\hbar\Omega_{q\lambda}$.

For a ferromagnet, we obtain from expression (5.14) and (5.12), taking (5.87), (5.90) and (5.93) into account,

$$\frac{1}{T_1} = \frac{\pi\hbar (\gamma_n H_n)^2}{4V_\rho} \sum_{q\lambda} \frac{|\omega_{ms}(e_z q_\pm + e_\pm q_z) + \omega_a (e_z q_\pm - e_\pm q_z)|^2}{\omega_q^2 \Omega_{q\lambda}} \times$$

$$\times (1 + 2\bar{N}_{q\lambda}) \delta(\Omega_{q\lambda} - \omega_n). \qquad (5.94)$$

Averaging over the polarization $e_{q\lambda}$ and integrating over q for $\omega_q \simeq \omega_e$ (for very small values of q, when $\Omega_{q\lambda} = \omega_n$) and $\kappa T \gg \hbar\omega_n$, we finally have

$$\frac{1}{T_1} = \frac{(\gamma_n H_n)^2 \omega_n^2 (\omega_{ms}^2 + \omega_a^2) \kappa T}{3\pi\rho c_s^5 \omega_e^2}. \qquad (5.95)$$

Expression (5.95) is applicable to the nuclei of both magnetic and nonmagnetic atoms. In the magnetic case $\omega_n \simeq \gamma_n H_n$. Accordingly the characteristic temperature and field dependence of the spin lattice relaxation time (for $H \gg H_a$, M_s, when one can neglect in ω_e the anisotropy and demagnetization fields) is of the form $T_1 \sim H^2/T$. For usual ferromagnets, however, the magnitude of T_1 given by (5.95) is many orders greater than the experimental values. Even for the optimal values of the parameters $\omega_{ms} \sim 10^{13}$, $\omega_n \sim 10^9$, and $\omega_e \sim 10^{10}$ sec^{-1}, assuming $\rho \sim 5$ g/cm^3 and $c \sim 2 \cdot 10^5$ cm/sec, we obtain from expression (5.95) $T_1 \sim 10^2/T$ (sec). It is probably this relaxation mechanism that plays a significant role in the dielectric ferromagnet

EuS, for which at temperatures $T \sim 0.1°K$ the relaxation time $T \sim 10^4$ sec (Schermer and Passel, 1965; Honma, 1966).

In the case of an antiferromagnet of the EA type, expression (5.88) for $\delta H^{\pm}$ is obtained from the analogous expression (5.87) for a ferromagnet by the substitution $\omega_k^{-1} \rightarrow |\gamma_e| H_E \omega_k^{-2}$. But for small k, which are of importance in relaxation processes, $\omega_k^2 \sim \gamma_e^2 H_E H_a$ for an antiferromagnet of the EA type. As a result, for these ferromagnets $1/T_1$ will again be determined by expression (5.95) if we replace ω_e by $\omega_a = |\gamma_e| H_a$. In the first approximation $1/T_1$ will not depend on H. The linear temperature dependence of $1/T_1$ at low temperatures was observed in the classical antiferromagnets MnF_2 (Kaplan et al., 1966) and $CuCl_2 \cdot 2H_2O$ (Benoit and Renard, 1964) for F^{19} nuclei and protons, respectively.

However, for a quantitative agreement with the theory, very large values for the parameter ω_{ms} again have to be taken.

The spin-lattice relaxation caused by the above mechanism should be highly effective in the case of an antiferromagnet of the EP type, which has a low-frequency spin vibration branch with a frequency ω_{1k}. In this case, instead of (5.95) we obtain (Turov and Timofeev, 1967)

$$\frac{1}{T_1} = \left(\frac{\gamma_e H_E}{\omega_{e1}}\right)^2 \frac{(\gamma_n H_n^{\zeta})^2 (\omega_{ms}^{(1)})^2 \kappa T}{6\pi\rho c_s^5 \omega_{e1}^2}. \qquad (5.96)$$

The last expression differs from the corresponding expression (5.95) for a ferromagnet only in that it contains a large multiplier* $(\gamma_e H_E/\omega_{e1})^2 \sim 10^4$–$10^6$. Here only the part of $1/T_1$ associated with the first term in expression (5.89) for $\delta H^{\pm}$ was taken into consideration. The second term, associated with the interaction between the nuclei and the lattice through the second (high-frequency) spin wave branch, makes a contribution of the same order to $1/T_1$ as in an antiferromagnet of the EA type. Note that for high fields, expression (5.96) leads to a dependence of the form $T_1 \sim H^4/T$.

Unfortunately, expression (5.96) contains unknown parameters (ω_{ms} and c_s) which, for antiferromagnets of the EP type ($MnCO_3$, $CsMnF_3$, etc), do not allow an accurate assessment of the contribution of this mechanism to the longitudinal relaxation rate. Almost no experimental data are available on T_1 for these antiferromagnets. In the only work known to us, that of Welsh (1966), a value of $T_1 = 3.7$ sec was obtained for the Mn^{55} nuclei in $CsMnF_3$ at $T = 1.4°K$ and $H = 5000$ Oe, T_1 varying with temperature approximately as T^{-5}.

From the point of view of theoretical estimates, the situation is somewhat better for the Fe^{57} nuclei in the compound α-Fe_2O_3 (hematite) which (above 260°K) is an

---

* For an antiferromagnet of the EA type, the analogous multiplier, as we have mentioned above, cancels out because of the high value of the two AFMR frequencies: $\omega_e \simeq \sqrt{\omega_E \omega_a}$. For antiferromagnets of the EP type, this does not take place because of the presence of low-frequency AFMR branches.

antiferromagnet of the EP type with weak ferromagnetism. The magneto-elastic parameter $\omega_{ms}^{(1)}$ for hematite can be estimated from the magnetostrictive AFMR frequency shift (Borovik-Romanov and Rudashevskii, 1964). This gives $\omega_{ms}^{(1)} = \gamma_e \hbar G_1/M_0 \simeq 2 \cdot 10^{12}$ sec^{-1}. If we also take, in accordance with known data, $H_E = 2 \cdot 10^6$ Oe, $\omega_{e1} = 2 \cdot 10^{10}$ sec^{-1} (as $H \to 0$), $\omega_n = 4 \cdot 10^8$ sec^{-1}, $\rho = 5$ g/cm^3 and $c_s = 4.7 \cdot 10^5$ cm/sec, then at $T = 300°$K we have from (5.96) $T_1 \sim 10^{-1}$ sec. This is comparable with the experimentally observed value $T_1 \sim 10^{-3} - 10^{-1}$ sec found in hematite at room temperature (Matsuura et al., 1962).*

Note that expression (5.96), in general, is not applicable to very low temperatures, when the correlation effects caused by the dynamic coupling of electron and nuclear spins (Chapter II, Sec. 3) become important. This does not refer to the case of a very low concentration of magnetic nuclei, when the correlation effects are absent.

Finally, we shall briefly consider one-phonon relaxation processes associated with the modulation of the anisotropic HFI, which are derived from the energy of spin-lattice interaction of the form (5.82) or (5.83). Taking again $\delta H^{\pm}$ according to (5.86), we can calculate $1/T_1$ from (5.14) and (5.12). We thus find, for both ferromagnets and antiferromagnets, an expression which, apart from a numerical factor, coincides with that obtained from (5.95) when we take $\omega_{ms}^2 + \omega_a^2 \equiv \omega_e^2$. Instead of H_n it contains a constant which is only determined by the anisotropic part of HFI. For example, in the case (5.83),

$$H_n^2 \to [(A_{\perp} - A_z) \langle S_i^z \rangle / \gamma_n \hbar]^2 \tag{5.97}$$

(for cylindrical symmetry of the HFI tensor, when $A_x = A_y = A_{\perp}$). For (5.82),

$$(\gamma_n \hbar H_n)^2 \to \left| \sum_{j\alpha\beta} \frac{\partial A_{xz}}{\partial R_j^{\alpha}} R_j^{\beta} \langle S_j^z \rangle \right|^2 + \left| \sum_{j\alpha\beta} \frac{\partial A_{yz}}{\partial R_j^{\alpha}} R_j^{\beta} \langle S_j^z \rangle \right|^2 ; \tag{5.98}$$

the sum is taken over magnetic atoms closest to the nucleus under consideration.**

---

* These authors obtained T_1 between $3 \cdot 10^{-3}$ and 10^{-1} sec, depending on the intensity of the rf pulses, with the upper limit corresponding to high intensities ($h \sim 1$ Oe). The lower limit presumably corresponds to nuclear spin relaxation in the domain walls, while the upper limit, as a consequence of NMR saturation in domain walls, is characteristic to some extent of nuclear spin relaxation inside the domains as discussed by us.

** Relation (5.98) takes into account also the "discrete" part of the dipole field (2.4) (due to the magnetic moments lying inside the Lorentz sphere). As regards the dipole contribution of magnetic moments lying outside the Lorentz sphere, its modulation by lattice vibrations can be taken into account in accordance with (5.50) if instead of the local magnetization $M(r)$ per unit volume we introduce the magnetization $\mathfrak{M}(r) = M(r)/\rho$ per unit mass and the density ρ is then expanded in a series in powers of the strain,

$$\rho = \rho_0(1 - u_{\alpha\alpha}).$$

For a ferromagnet, the term with $\partial M_z/\partial z$ in (5.50) then gives for $M_z \cong M_0(\rho/\rho_0)$ the local field

$$\delta H = - 4\pi M_0 \sum_k (k_z/k^2) k u_{\alpha\alpha}(k),$$

which leads to an expression for $1/T_1$ by replacing in (5.95) $(\omega_{ms}\gamma_n H_n/\omega_e)$ by $(\gamma_n \hbar/10) 4\pi M_0$.

Phonon–Magnon or Two-Phonon Processes

One-phonon relaxation processes usually prove ineffective because of the small density of states for phonons with frequency $\Omega_{q\lambda} = \omega_n$, which take part in these processes. It is, therefore, of interest to consider processes in which several quasi-particles take part, for example, a phonon and a magnon or two phonons.

Processes of the first type can be produced by HFI modulation, if now we take into account in (5.82) or (5.83) terms with transverse components of the electron spins S_j^x and S_j^y. Then $\delta H^{\pm}$ will be proportional to products of $u_{\alpha\beta}$ and $S^{\pm}$, and, in the second quantization representation, to products of phonon and magnon creation and destruction operators. Because of the energy conservation law, only terms with products of the form $b_{q\lambda}^{+} a_k$ and $b_{q\lambda} a_k^{+}$ will be important: these terms correspond to processes of phonon creation and magnon destruction (or vice versa) for which $\Omega_{q\lambda} - \omega_k \pm \omega_n = 0$.

Note that these processes, if they are caused by the interaction (5.82), will also take place for isotropic HFI, when $A^{\alpha\beta} = A\delta_{\alpha\beta}$.

Let us present the final results for $1/T_1$ in this case, omitting again simple calculation based on expressions (5.14) and (5.12). For the temperature interval where

$$T \gg \hbar\omega_e/\kappa, \quad T \ll T_C(T_N) \quad \text{and} \quad T \ll T_D$$

(T_D is the Debye temperature) we have for a ferromagnet and an antiferromagnet, respectively,

$$\text{(F)} \qquad \frac{1}{T_1} \simeq 10^{-2} \frac{\hbar}{Ma^2} \frac{(\Delta A)^2}{\hbar^2 \omega_E^{3/2} \Omega_s^{1/2}} \left(\frac{\kappa T}{\hbar\Omega_s}\right)^{9/2},$$

$$(\Delta A)^2 = \sum_{\alpha\beta} \left(\sum_j \frac{\partial A}{\partial R_j^{\alpha}} R_j^{\beta}\right)^2 ; \tag{5.99}$$

$$\text{(AF)} \qquad \frac{1}{T_1} \simeq 10^{-2} \frac{\hbar}{Ma^2} \frac{(\Delta A)^2}{\hbar^2 \omega_E^2} \left(\frac{\kappa T}{\hbar\Omega_s}\right)^5,$$

$$(\Delta A)^2 = \sum_{\alpha\beta} \left(\sum_{j_1} \frac{\partial A}{\partial R_{j_1}^{\alpha}} R_{j_1}^{\beta} - \sum_{j_2} \frac{\partial A}{\partial R_{j_2}^{\alpha}} R_{j_2}^{\beta}\right)^2, \tag{5.100}$$

where M is the mass in a unit cell, a is the linear dimension of the unit cell, $\Omega_s = c_s/a$, and ΔA is a parameter characterizing the change of HFI under strain; R_j is the radius vector of the magnetic atoms closest to the nuclear spin under consideration (in the case of an antiferromagnet, j_1 and j_2 identify the sites of the first and the second magnetic sublattices respectively). For a rough order-of-magnitude estimate, it can be assumed that $\Delta A \sim A$.

The experimental data for the Mn^{55} nuclei in the antiferromagnet $CsMnF_3$ (Welsh, 1966, 1967) agree with the $T_1 \sim T^{-5}$ temperature dependence predicted by (5.100). It is impossible to carry out a quantitative comparison because the accurate values of the parameters ΔA, ω_E, and c_s are not known. Nevertheless, the experimental value $T_1 = 3.7$ sec at $1.4°K$ can be obtained from expression (5.100) for permissible values of these parameters.*

Expressions (5.99) and (5.100) apply to the temperature interval below the Debye temperature T_D. If $T_{C,N} \gg T_D$, another limiting case $T_{C,N} \gg T > T_D$ may be considered. Accordingly, we again obtain expressions of the form (5.99) and (5.100) with the only difference that the powers 9/2 and 5 of the last factor (which contains T) are replaced in both expressions by 2, so that in this case $T_1 \sim T^{-2}$ for both ferromagnets and antiferromagnets.

This temperature dependence mentioned ($T_1 \sim T^{-9/2}$ or T^{-5} for $T < T_D$ and $T_1 \sim T^{-2}$ for $T > T_D$) is retained for other spin-lattice interaction mechanisms, which lead to scattering processes with the participation of a magnon and a phonon. We are referring to the interaction described by the Hamiltonian (5.83) and also the interaction due to the modulation of the dipole field (5.50) by density oscillations $\rho_0 - \rho = \rho_0 u_{\alpha\alpha}$ for the terms $\partial M_x/\partial x$ and $\partial M_y/\partial y$ in div M ($M_{x,y} = \rho \mathfrak{M}_{x,y}$).

Finally, if we take into account quadratic terms in the expansion of the hyperfine or dipole interactions in powers of strain, we find that two-phonon Raman scattering processes are possible. As in the case of paramagnets, the spin-lattice relaxation rate will be proportional to T^7 or T^2 at low ($T < T_D$) or high ($T > T_D$) temperatures, respectively (see, for example, Abragam, 1963).

The presence of an indirect interaction between nuclear spins and lattice vibrations via spin waves leads to a two-phonon relaxation mechanism which is specific to magnetically-ordered crystals. In the effective Hamiltonian $\mathcal{H}_{IL}$ there will appear quadratic terms with respect to strain if this interaction is calculated in a higher order of perturbation theory and if terms of second order in the magneto-striction constant are taken into account. For low temperatures, we obtain again $T_1 \sim T^{-7}$. The appropriate expression is given by Pincus and Winter (1963).

A temperature dependence close to T^{-7} was observed for the spin-lattice relaxation rate of protons in the antiferromagnetic compounds $CuCl_2 \cdot 2H_2O$ (Benoit and Renard, 1964) and $CoCl \cdot 6H_2O$ (Abkowitz and Lowe, 1966). It is not clear as yet which of the two-phonon relaxation mechanisms plays the main role here.

---

* We have already mentioned that three-magnon scattering processes also lead to a $T_1 \sim T^{-5}$ temperature dependence and Welsh (1966, 1967) explained the experimental data in terms of these processes (see also Beeman and Pincus, 1968).

8. Some Conclusions

The above discussion relating to the NMR line broadening and nuclear spin-lattice relaxation mechanisms specific to ferro- and antiferromagnets shows that there is a large variety of such mechanisms associated in the main with the active participation of spin waves in NMR phenomena in magnetically-ordered crystals. Today, because of the paucity of experimental data (especially for single-domain specimens), it is often impossible to tell with any accuracy which is the decisive mechanism for a particular magnetic crystal. Nevertheless, a number of mechanisms can be distinguished which, at least, in individual cases, play an important role.

In NMR line broadening at a sufficiently high concentration of magnetic nuclei, the Suhl–Nakamura mechanism of nuclear spin interaction via spin waves is the most important. It leads, in the first approximation, to a line width $\Delta\omega_n$ which is independent of temperature. The marked temperature dependence of $\Delta\omega_n$ for some antiferromagnets of the EP type—the manganese compounds ($MnCO_3$, $CsMnF_3$, etc.)—in the region of liquid helium temperatures is apparently caused by correlation effects. For these antiferromagnets, nonuniform broadening caused by the nonuniform NMR dynamic frequency shift, as a result of coupling between the electron and nuclear spin vibrations (if the AFMR has a substantial nonuniform broadening) also makes a temperature-dependent contribution to $\Delta\omega_n$.

At low magnetic isotope concentrations, there is apparently no universal mechanism of NMR line broadening. It is clear that in dielectrics, line broadening is due to various processes of spin wave scattering by nuclear spins. In terms of free spin waves (magnons), these are one-magnon processes in the case when there is a strong damping of longitudinal spin waves at the NMR frequency (e.g., due to their interaction with fast relaxing magnetic impurities in the case of the Fe^{57} nuclei in yttrium ferrite garnet at temperatures $T < 20°K$). The same mechanism will also serve as a cause of spin-lattice relaxation. In other cases, two-magnon scattering processes can be the cause of broadening (again, the yttrium garnet at $T > 20°K$ can serve as an example). If the HFI is anisotropic, the same processes may cause longitudinal relaxation. Scattering with the participation of a large number of magnons also deserves attention.

In metallic ferromagnets, the transverse and longitudinal relaxations are apparently produced by the fluctuations of the orbital and contact fields set up by the collective electrons. A linear temperature dependence is characteristic for the two relaxation rates.*

Finally, various mechanisms of interaction between nuclear spins and lattice

---

* Note that in the case of metals it is generally necessary to take into account electromagnetic effects caused by the high electric conductivity and the skin-effect (Buishvili et al., 1964).

vibrations may contribute to spin-lattice relaxation in dielectrics. In crystals with very large magnetostriction, these may be one- and two-phonon processes produced by the indirect interaction between the nuclei and the lattice via spin waves. It is possible that in some cases complicated relaxation processes with the participation of magnons and phonons, specific for magnetic crystals, play a role.

The majority of the above propositions are only approximate. For the final solution of the problem of NMR line width and relaxation further experimental investigations are needed on relaxation phenomena and their dependence on temperature and external fields. Nuclear magneto-acoustic resonance can assist in the clarification of the role of various spin-lattice coupling mechanisms.

CHAPTER VI

Quadrupole Effects

1. The Quadrupole Hamiltonian

If the nuclear spin $I > \frac{1}{2}$, the nucleus possesses an electric quadrupole moment and therefore the inhomogeneous internal electric fields will affect the NMR. In particular, if the symmetry of the electric field gradient tensor $V_{\alpha\beta}$ acting at the nucleus is lower than cubic, the NMR line is split into $2I$ components.

In a coordinate system X, Y, Z associated with the main axes of this tensor, the effective quadrupole Hamiltonian has the form (Abragam, 1963, Chapter VIII)

$$\mathscr{H}_Q = \frac{1}{2}\hbar\omega_Q [I_Z^2 - \frac{1}{3}I(I + 1) + \frac{1}{6}\eta(I_+^2 + I_-^2)], \tag{6.1}$$

$\hbar\omega_Q = 3e^2qQ/2I(2I - 1)$, Q is the quadrupole moment of the nucleus,

$$eq = V_{ZZ}, \quad \eta = \frac{V_{XX} - V_{YY}}{V_{ZZ}}.$$

The electric field gradient at the nuclei and the atoms of transition elements is mainly produced by the unfilled d- or f-shell electrons. Therefore the main axes of the gradient tensor $V_{\alpha\beta}$ and the HFI tensor usually coincide. However, the quantization axis for nuclear spins (directed along the local magnetic field) does not necessarily coincide with any of these axes, since the equilibrium orientation of electron spins, which also basically determines the direction of the local field at the nucleus, is established by a number of competing forces of different nature: exchange forces, magnetic anisotropy, and the external magnetic field. This circumstance must be borne in mind especially for nuclei situated in domain walls of ferromagnets and antiferromagnets, and also in the case of nuclear resonance in complex noncolinear (e.g., helical) magnetic structures.

Let the tensor $V_{\alpha\beta}$ have axial symmetry ($\eta = 0$), and let the quantization axis z of nuclear spins make an angle θ with the symmetry axis Z. It can be assumed, without loss of generality, that the coordinate system XYZ associated with the main axes of the tensor $V_{\alpha\beta}$ can be obtained from the coordinate system xyz associated with the spin quantization axis by a rotation through an angle θ about the common axis $x \equiv X$. The full effective nuclear spin-Hamiltonian, taking HFI and quadrupole interactions into account, will then have the following form in the xyz system:

$$\mathscr{H} = - \hbar\omega_n I_z + \tfrac{1}{2}\hbar\omega_Q \{(1 - \tfrac{3}{2}\sin^2\theta)[I_z^2 - \tfrac{1}{3}I(I+1)] +$$

$$+ \tfrac{1}{4}\sin^2\theta(I_+^2 + I_-^2) - \tfrac{1}{4}\sin 2\theta[I_z(I_+ - I_-) + (I_+ - I_-)I_z]\}, \qquad (6.2)$$

where $I_\pm = I_x \pm iI_y$.

For nuclei in magnetic atoms we usually have $\omega_n \sim \omega_{n0} \gg \omega_Q$, therefore the quadrupole interaction can be considered as a small perturbation (in comparison with the quasi-Zeeman hyperfine interaction). In first-order perturbation theory, the eigenvalue of the Hamiltonian $\mathscr{H}$ (6.2) is determined by the expression (Abragam, 1963)

$$E_m = - \hbar\omega_n m + \tfrac{1}{2}\hbar\omega_Q[m^2 - \tfrac{1}{3}I(I+1)](1 - \tfrac{3}{2}\sin^2\theta) \qquad (6.3)$$

(m is the magnetic quantum number, having $2I + 1$ values between $-I$ and $+I$). The corresponding wave functions for a given m contain an admixture (of the order of ω_Q/ω_n) of states with $m \pm 1$ and $m \pm 2$.

Below we shall only limit ourselves to a simple enumeration of the basic effects to which the quadrupole changes of the energy spectrum and the wave functions lead.

2. Quadrupole Splitting and Broadening of the NMR Line

We shall first consider one feature of the splitting of the NMR line into $2I$ components that has already been mentioned. As is seen from (6.3), this splitting for a half integer spin in the first order perturbation theory is such that, in addition to a central line corresponding to a transition between levels with $m = +\tfrac{1}{2}$ and $m' = -\tfrac{1}{2}$ which remains unshifted, there also appear $2I - 1$ symmetrically arranged satellites.*
Quadrupole splitting was observed for Mn^{55} nuclei ($I = \tfrac{5}{2}$) in Mn_3O_4 (Houston and Heeger, 1966), in Mn_4N (Hihara et al., 1963; Matsuura, 1966) and in Mn_5Ge_3 (Hihara et al., 1963), Cr^{53} nuclei ($I = \tfrac{3}{2}$) in $CrBr_3$ (Gossard et al., 1962) and in Cr_2O_3 (Rubinstein et al., 1964), Eu^{153} nuclei in EuS (Boyd, 1963).

---

* The "mixing" of states with different magnetic quantum numbers m, caused by the quadrupole interaction, "allows" forbidden transitions with the selection rules $\Delta m = \pm 2$ and ± 3. However, the intensities of these lines are proportional to $(\omega_Q/\omega_n)^2$ and, as a rule, are small in comparison with the intensity of the fundamental transition lines $\Delta m = \pm 1$ (Buishvili, 1963a, b, c).

If there is a nonuniform spread in the values of the angle θ, a broadening of the satellite lines is observed (quadrupole broadening of the central component takes place only in the second-order perturbation theory). In materials with magnetic order, the nonuniformity is mainly associated with nonuniformity in the directions of the electron spins (which leads to a nonuniform spread in the directions of the local magnetic fields). When the spread in θ is sufficiently large ($|\Delta \sin \theta| \sim 1$), the broadening of the satellite lines will be comparable with the line separations. Accordingly, the satellites will not be observed and their existence will be evident only in the reduction of the intensity of the central line. Such a situation obtains, for example, for nuclei in domain walls: the direction of the local field in the transition layer between two domains varies in a quasi-continuous way from one nucleus to the next (see Chapter V, Sec. 7).

Various crystal lattice defects, changing the symmetry or the magnitude of the electric field gradient at the nuclei (in cubic crystals as well), can have a substantial influence on line widths and shapes, as in nonmagnetic crystals (Abragam, 1963).

The following two effects are specific to ferro- and antiferromagnets.

The first effect is connected with the interaction between the individual nuclear spin and the fluctuations of the local fields. As we have seen in the previous chapter, when $I \neq \frac{1}{2}$ this interaction not only shifts and broadens the NMR line, but also leads to its splitting, which is analogous to the quadrupole splitting considered here. It is characteristic, however, that the splitting contained in (5.26) will occur no matter how high the symmetry of the surroundings of the nuclear spin. This refers to crystallographic symmetry, since the magnetic symmetry will not be higher than cylindrical (axial) due to the presence of a preferred ferro- or antiferromagnetic axis.

This effect has, in fact, already been derived in Chapter IV in the free spin wave approximation. Virtual processes of emission and absorption of spin waves by the nucleus, caused by the HFI, lead to the appearance of an axisymmetrical quadrupole Hamiltonian even in cubic crystals. In particular, for a ferromagnet, this Hamiltonian has the form

$$\delta \mathcal{H}_Q = \tfrac{1}{2} \hbar \delta \omega_Q I_z^2 , \qquad (6.4)$$

where

$$\hbar \delta \omega_Q = -\frac{SA^2}{N} \sum_k \frac{1}{\hbar \omega_k} .$$

Thus, this interaction can lead to quadrupole splitting of the NMR line in cubic crystals, the relative magnitude of the splitting being

$$\frac{\Delta \omega_Q}{\omega_{n0}} = \frac{1}{N} \sum_k \frac{A}{\hbar \omega_k} \sim \frac{A}{J} ,$$

where J is the exchange interaction parameter. Usually $A/J \sim 10^{-4}$ and less; therefore the splitting apparently cannot be observed, being less than or of the order of the

width of the NMR line. It is, however, possible that in some cases, e.g., for rare earths in compounds with low Curie (Néel) points, this splitting will be substantial.

The second effect is caused by the correlation of the nuclear spin motions (at a sufficiently high concentration of magnetic nuclei) because of their indirect interaction via spin waves. As we saw in Chapter II, the coupling of electron and nuclear spin vibrations brought about by HFI leads to a dynamic shift $\delta\omega_n$ in the NMR frequency, which increases with the lowering of the temperature, while the static polarization of the nuclear spin system $\langle I_z \rangle$ increases. This shift is due to the fact that the nuclear spins, interacting via spin waves, vibrate in a consistent manner in some correlation region. As shown by Kurkin and Parfenova (1966), this correlation can significantly affect the quadrupole splitting, if such a splitting exists. The magnitude and character of the effect depends on the ratio between the dynamic shift parameter $\delta\omega_n$ and the quadrupole interaction parameter ω_Q.

At sufficiently high temperatures (remember that $\delta\omega_n$ decreases with increasing temperature as $1/T$), when $\delta\omega_n \ll \omega_Q$, the quadrupole splitting is preserved, but the components of the quadrupole multiplet on the lower frequency side are shifted by an amount proportional to $\delta\omega_n$. The symmetrical arrangement of the satellites about the central line is broken. For example, in a ferromagnet with $I = \frac{3}{2}$ we have three lines:
$$\omega_1 = \gamma_n(|H_n + H|) - 0.2\delta\omega_n, \quad \omega_{2,3} = \gamma_n(|H_n + H|) - 0.15\delta\omega_n \pm \omega_Q.$$

With the lowering of temperature, as the role of correlation effects increases, the quadrupole splitting first decreases and then, for $\delta\omega_n \gg \omega_Q$, it vanishes altogether (in the linear approximation in ω_Q). We obtain one resonance line shifted by $\delta\omega_n$:
$$\omega_{1,2,3} = \gamma_n |H_n + H| - \delta\omega_n.$$

The most favorable conditions for the observation of the effect of disappearance of quadrupole splitting with the lowering of temperature can be expected in anti-ferromagnets of the EP type, for which $\delta\omega_n/\omega_n \sim \omega_T^2/\omega_e^2$ can attain tens of percent already in the range of helium temperatures (see Chapter II, Sec. 3). Of course, the quadrupole interaction (the parameter ω_Q) should also be sufficiently large, certainly larger than the NMR line width.

In conclusion, let us emphasize that the quadrupole splitting of the NMR line, as a result of correlations in the motion of nuclear spins, will vanish long before magnetic ordering takes place in the nuclear spin-system.

3. Effect on Relaxation Processes

Quadrupole interaction may also affect the relaxation times T_2 and T_1 associated with the fluctuations of the local hyperfine field at the nucleus (see Chapter V, Sec. 5).

Appropriate calculations carried out by Moriya (1956) show that quadrupole interaction has a weak effect on $1/T_2$, introducing a factor of the form $[1 + \text{const} \times \times (\omega_Q/\omega_n)^2] \sim 1$ in (5.22). Yet the influence of $\mathcal{H}_Q$ on the thermal relaxation

rate $1/T_1$ can be substantial when the spin quantization axis (z) does not coincide with the electric field gradient axis (X, Y or Z). While for $\mathscr{H}_Q = 0$ the mean probability of the thermal transition $W_{m \to m \pm 1} \sim 1/T_1$ vanishes in the first approximation, the situation is different for $\mathscr{H}_Q \neq 0$. First, transitions between any two levels m and m' are allowed. Second, all the transitions can be divided into two groups: 1) transitions without a change in the absolute value of m (for example $\frac{1}{2} \to -\frac{1}{2}$) and 2) transitions with a change in $|m|$ (for example, $\frac{1}{2} \to \pm \frac{3}{2}$). The probability of a thermal transition of the first type ($|m'| = |m|$) is

$$W_{m \to m'} \sim \frac{1}{T_2} \left(\frac{\omega_Q}{\omega_n} \right)^6$$

while the probability for transitions of the second type ($|m'| \neq |m|$) is

$$W_{m \to m'} \sim \frac{1}{T_2} \left(\frac{\omega_Q}{\omega_n} \right)^2 .$$

Therefore, when $\omega_Q \ll \omega_n$, the spin-lattice relaxation rate $1/T_1$ for the fundamental NMR line ($\frac{1}{2} \to -\frac{1}{2}$) may be determined by thermal transitions via intermediate states (transitions $\frac{1}{2} \leftrightarrow \pm \frac{3}{2} \leftrightarrow -\frac{1}{2}$) and not by direct relaxation ($\frac{1}{2} \leftrightarrow -\frac{1}{2}$).

Finally, modulation of the quadrupole interaction by lattice vibrations leads to yet another spin-lattice coupling mechanism (in addition to that considered in Chapter V, Sec. 7). The cubic symmetry of the electric field gradient is broken by lattice vibrations, even in cubic crystals, so that the quadrupole mechanism of spin-lattice coupling will be effective in these crystals. The linear terms in the expansion of the quadrupole Hamiltonian in powers of the strain will cause one-phonon processes of spin-lattice relaxation (with a temperature dependence of the form $1/T_1 \sim T$), while the quadratic terms will cause two-phonon Raman processes (with a temperature dependence of the form $1/T_1 \sim T^7$ or $1/T_1 \sim T^2$ at low or high temperatures, respectively). The calculation of the longitudinal relaxation rate for this mechanism in fact does not depend on whether the crystal is magnetic or not (see Abragam, 1963).

It should, however, be pointed out that the strain mechanism of quadrupole interaction modulation is the only mechanism considered in the current literature (the expansion is in powers of the components of the tensor $u_{\alpha\beta}$). It is clear, by analogy with the magneto-elastic interaction (see Chapter V, Sec. 7), that the rotation of the main axes of the tensor $V_{\alpha\beta}$ (together with a volume element) will introduce additional terms in the Hamiltonian of the spin-lattice quadrupole interaction, which are proportional to the components of the tensor $\varepsilon_{\alpha\beta}$. For example, for uniaxial symmetry

$$\mathscr{H}_{IL}^{(\varepsilon)} \equiv \Delta \mathscr{H}_Q^{(\varepsilon)} = \hbar \omega_Q (I_z I_x \varepsilon_{zx} + I_z I_y \varepsilon_{zy}).$$

For a quantitative comparison of theory with experiment, these terms must be taken into account if the quadrupole mechanism of spin-lattice coupling contributes substantially to $1/T_1$. All this applies to magnetic as well as nonmagnetic (or paramagnetic) crystals.

CHAPTER VII

Nuclear Magnetic Resonance in Domain Walls

1. Domain Structure and NMR

Most of the experimental work on NMR in ferromagnets has been confined until recently to the observation of resonance for nuclei situated in domain walls. The NMR signal from these nuclei is appreciably larger than that originating from nuclei inside the domains (or in single-domain specimens), and the investigation of resonance for nuclei in domain walls does not require very sensitive apparatus. However, the theoretical treatment of NMR in the presence of domain structure is substantially more difficult than in single-domain specimens. Therefore, experiments in single-domain specimens are, from the theoretical point of view, in general preferable.

Nevertheless, NMR in multi-domain ferro- and antiferromagnets is of considerable interest as it provides one of the most powerful investigation methods of domain structure. Therefore we will briefly discuss the characteristic features of NMR associated with the domain structure (basically in relation to nuclei situated in transition layers between domains, i.e., domain walls, where these features are the most prominent). There is no rigorous quantitative theory of NMR in multi-domain specimens, and we shall only present the basic ideas explaining in the main the qualitative aspects of the phenomena. We shall limit ourselves to ferromagnets.

The main feature, i.e., the substantially large amplification coefficient η for nuclei in the walls in comparison with nuclei in domains, has already been mentioned in Chapter I. Although a more detailed discussion of various aspects of NMR intensity in domain walls is possible, we shall not dwell on these details and refer the reader to original work.*

* Gossard and Portis (1959), De Gennes et al. (1963), Herve and Aubrun (1962), Robert (1962), Buishvili and Giorgadze (1963), Murray and Marshall (1965), Onoprienko (1966).

A second group of features are associated with the nonhomogeneous distribution of magnetization in specimens with domain walls. The magnetization gradients have the largest value in the transition region between domains (see Figure 2). Since the NMR frequency is determined by the local magnetic field at the nucleus, we can approximately assume for nuclei of magnetic atoms

$$\omega_n = \gamma_n \left| H_n + H_{dip}^z \right|. \tag{7.1}$$

Here $H_n = - A_0 M$ is the hyperfine field at the nucleus oriented along (usually against) the local magnetization M, the direction of which is taken as the Z axis of the local coordinate system (Figure 14); H_{dip}^z is the projection of the dipole field H_{dip} on this axis. If the distribution of the magnetization in the sample is known, H_{dip} can be calculated in principle at each point inside the specimen; it consists of a continuous magnetostatic part H_{mst} and a discrete part H_{dis} so that $H_{dip} = H_{mst} + H_{dis}$. The field H_{mst} is the solution of the magnetostatic equations (curl $H = 0$, div $H = - 4\pi$ div M), being a field created by a known continuous distribution of volume and surface magnetic charges,

$$H_{mst} = - \nabla \left\{ \int_V \frac{- \text{div } M(r') \, dr'}{|r - r'|} + \int_\Sigma \frac{M(r') \, n dS'}{|r - r'|} \right\}. \tag{7.2}$$

In the approximation of a true continuum, the first integral is taken over the entire volume of the specimen, while the second is taken on its surface (n is the unit vector of the outer normal to the surface). As with single-domain specimens (see Chapter II, Sec. 1), the discrete character of the medium can be taken into account by identifying in the dipole sum (2.4) the contribution H_{dis} from the magnetic moments inside the Lorentz sphere (centered at the point of interest r). The surface integration in (7.2) is then carried out not only over the outer surface of the specimen (Σ) but also over the surface of the Lorentz sphere (σ).

Since the effective thickness of the wall is usually hundreds of interatomic distances, the radius of the Lorentz sphere may always be chosen much smaller than the wall thickness. It can, therefore, be assumed that the magnetic moments inside the Lorentz sphere are approximately parallel. As a result, the surface integral over σ gives a

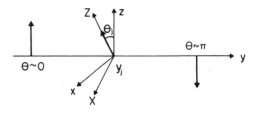

Figure 14

The local system of coordinates in a 180° domain wall.
The Z axis is everywhere oriented along the equilibrium direction of the spins.

Lorentz field $H_L = (4\pi/3) M(r)$ at the point r which is parallel to the local magnetization $M(r)$, so that it simply leads to a renormalization of the HFI constant A_0. The field H_{dis}, representing the sum of the dipole fields of the magnetic moments inside the Lorentz sphere, will vanish in this approximation (neglecting the non-parallel alignment of the spins in the Lorentz sphere) for nuclei with cubic surroundings and will depend on the direction of $M(r)$ for nuclei with surroundings of lower symmetry.

In the general case for arbitrary domain structure, the field H_{mst} (7.2) will be substantially nonhomogeneous. However, as is shown in the theory of the ferromagnetic phase (Landau and Lifshits, 1935; Collection, 1951), the distribution of magnetization for simple equilibrium domain structures tends to be such that no magnetic space charge is formed, so that div $M(r) = 0$. Moreover, for crystals with moderate magnetic anisotropy, the formation of so-called closure regions, for which the normal component of the magnetization of the sample also vanishes, is energetically favorable near the surface of the specimen. As a result, $H_{\mathrm{mst}} \simeq 0$ for such simple "classical" domain structures. In this case, H_{dis} makes the only significant contribution to the NMR frequency (7.1), but only for nuclei with noncubic surroundings. This may be the reason for the difference between the resonance frequencies of nuclei in domain walls and nuclei inside the domains (and also nuclei in differently oriented domains or walls). For example, in the model of a $180°$ wall shown in Figure 2, the magnetic moments inside the wall have components perpendicular to the c axis of the crystal, whereas inside the domains the magnetic moments are directed only along or against this axis. Therefore, due to the presence of the field H_{dis} and the dependence of its projection H^z_{dis} on the direction of M, the mean resonance frequency for nuclei in domain walls will differ from that for nuclei in the domains. In addition, the NMR line in domain walls must of course experience a certain broadening, since H^z_{dis} assumes a quasi-continuous range of values in the wall.

It must be borne in mind that for nuclei with noncubic surroundings, the HFI anisotropy can serve as another reason for the difference in the resonance frequencies in walls and domains. This will clearly be the case, for example, for the $180°$ wall in Figure 2, provided the constant A_0 takes on different values for $M \| c$ and for $M \perp c$.

$CrBr_3$ (hexagonal atomic surroundings—DO_2) is an example of a ferromagnet in which Gossard et al. (1962) observed the resonance of Cr^{52} nuclei simultaneously from domains and from walls, at two different frequencies. The difference in frequencies as a function of temperature is presented in Figure 15. The previously mentioned factors responsible for this difference (dipole field anisotropy and HFI) refer mainly to the part of the curve which is retained at temperatures $T \to 0° K$. The temperature-dependent part will be discussed below (Sec. 2).

Expression (7.1) ignores the influence of the external field on the magnetization process of a multi-domain ferromagnet. It is usually assumed that the influence of the external field is so small that no appreciable modification of domain structure

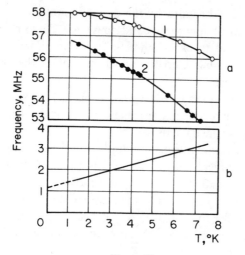

Figure 15

a The temperature dependence of the Cr^{53} NMR frequency in $CrBr_3$ (at $H = 0$) for nuclei in domains (curve 1) and in domain walls (curve 2); *b* the difference of NMR frequencies in domains and in walls as a function of temperature.

(or its disappearance) is observed, since its contribution to the total local field is cancelled by the demagnetization. There is, however, no theoretical treatment of this problem.

Finally, nor does expression (7.1) take into account the quadrupole splitting of the NMR frequency, which should occur for nuclei with $I > \frac{1}{2}$ in noncubic surroundings. For nuclei in domains, expression (7.1) remains nevertheless valid for the central component of the quadrupole multiplet (for half-integer I), which is not shifted in the first approximation. Since the magnitude of quadrupole splitting, according to (6.3), depends on the angle between $M(r)$ and the main axis of the electric field gradient tensor (in the axisymmetrical case), the quadrupole interaction for nuclei in a domain wall will only lead to an additional broadening of the NMR line. In $CrBr_3$, a triplet was observed for Cr^{53} ($I = \frac{3}{2}$) in domains and one broad line of a complicated shape for nuclei in domain walls.*

2. Spin Waves in Domain Walls

It has been shown for single-domain specimens that the various features of NMR in ferro- and antiferromagnets are essentially determined not only by the static dis-

---

* The intensity of this line was 30 times the integrated intensity of the triplet. This also provided an indication that the broad line was produced by nuclei in domain walls, whereas the narrower lines of the triplet were produced by nuclei in domains.

tribution of magnetization but also by the dynamic motion of the electron spins, namely by the spectrum of spin waves and spin wave damping and scattering by nuclear spins. It is natural to expect that the eigenvibrations of spins in domain walls will have their own characteristics which may affect the resonance of nuclei in multi-domain specimens.

Following the work of Winter (1961), we shall consider the simplest case of a 180° wall (Figures 2 and 14) in a uniaxial ferromagnetic crystal, where the easy axis coincides with the main symmetry axis, $c\|z$.*

The domain wall constitutes a transition layer between two plane domains, in one of which the spins are directed in the positive direction of the z axis, while in the other they point in the negative direction of the z axis. The direction perpendicular to the wall is taken as the y axis; we may then assume that S_j is parallel to z for $y_j \rightarrow -\infty$ and S_j is antiparallel to z for $y_j \rightarrow +\infty$. The angle θ_j fixing the equilibrium direction of a spin in the xz plane changes quasi-continuously from $\theta \sim 0$ (for $y \rightarrow -\infty$) to $\theta \sim \pi$ (for $y \rightarrow +\infty$).

The variation of θ_j as a function of y_j in a wall is determined by a competition between the exchange energy and the energy of magnetic anisotropy.** For our purpose, it is convenient to write the energy using the continuum model and replacing the discrete spins by a continuous magnetic moment density (local magnetization) $M(r)$. In this model, the exchange energy density (more precisely, that part which depends on the distribution of magnetization) can be presented in the form

$$\mathscr{H}_{ex}(r) = \frac{J_0 a^2}{M_0^2} [(\nabla M_x)^2 + (\nabla M_y)^2 + (\nabla M_z)^2], \tag{7.3}$$

where J_0 is the exchange interaction parameter having the dimensions of energy density. We shall write the magnetic anisotropy energy density in the form

$$\mathscr{H}_{an}(r) = \frac{K}{M_0^2} (M_x^2 + M_y^2). \tag{7.4}$$

If the distribution of equilibrium magnetization is found from the condition of minimum total energy

$$\mathscr{H} = \int (\mathscr{H}_{ex} + \mathscr{H}_{an}) dV \tag{7.5}$$

with the boundary conditions

$$M \rightarrow \pm M_0 \| z \quad \text{for} \quad y \rightarrow \mp \infty,$$

---

* See also the work of Janak (1964). The analogous problem of spin vibrations in domain walls of anti-ferromagnets was considered by Paul (1962) and Farztdinov (1966).

** The partition of the ferromagnet into domains, as is known, is associated with the requirement that the magnetostatic energy of the whole sample should be minimum (this is the result of surface magnetic charges).

we obtain the standard relation for the change of the magnetization rotation angle $\theta(y)$ in a domain wall,*

$$\frac{d\theta}{dy} = \frac{1}{\delta}\sin\theta. \qquad (7.6)$$

Here

$$\delta = a\sqrt{\frac{J_0}{K}} \qquad (7.7)$$

may be treated as the thickness of the domain wall. Indeed, the solution of the last equation with the above boundary conditions has the form

$$\sin\theta(y) = 1/\cosh\left(\frac{y}{\delta}\right), \qquad (7.8)$$

so that $\sin\theta \to 0$ when $|y| \gg \delta$.

We can then consider small eigenvibrations of the magnetization about a state with such a spatially nonhomogeneous (along the y axis) distribution.

Note, however, that the exchange ($\mathcal{H}_{ex}$) and the anisotropy ($\mathcal{H}_{an}$) energies as such determine only the angular distribution of the magnetization in the wall (with respect to the angle θ) and the wall thickness δ, but in no way do they fix the position of the wall (its center) in the coordinate y. Any arbitrarily small homogeneous perturbation may displace the wall as a whole to infinity. The existence of a domain wall and the finite value of the initial susceptibility of the ferromagnet indicate, however, that the wall is situated in a potential well. The energy needed to displace the wall to a distance Δy is equal to

$$\tfrac{1}{2}\alpha(\Delta y)^2, \qquad (7.9)$$

where α is the so-called coefficient of the quasi-elastic force (or the "rigidity") of the wall, which determines the initial (reversible) susceptibility of the ferromagnet. The displacement of the wall by Δy is equivalent, according to (7.6), to a rotation of the local magnetization through an angle $\Delta\theta = (\Delta y/\delta)\sin\theta$.

We introduce a new system of coordinates X, y and Z, taking the Z axis as the equilibrium direction of the magnetization for the original (unshifted) position of the wall (i.e., when $\Delta y = 0$), while the y axis remains as before (see Figure 14). The displacement of the wall by Δy gives rise, in this local coordinate system, to a transverse magnetization component

$$M_X = M_0\Delta\theta = M_0(\Delta y/\delta)\sin\theta. \qquad (7.10)$$

---

* More precisely, characterizing the direction of magnetization by the polar and the azimuthal angles $\theta(y)$ and $\phi(y)$, we obtain, in addition to equation (7.6), another equation $\partial\phi/\partial y = 0$, i.e., $\phi = $ const. We take $\phi = 0$, which is justified by subsequent calculation of the magnetostatic energy. Since the magnetic moments do not lie in the plane of the wall, demagnetization fields arise at the walls (see below).

It is easy to see that the quasi-elastic wall energy (7.9) can be described in terms of some additional effective local anisotropy, with an energy density of the form

$$\mathcal{H}'_{an} = \frac{K'}{M_0^2} M_X^2 . \tag{7.11}$$

Indeed, $\int \mathcal{H}_{an} \, dV$, taking (7.10) into account, leads to an expression of the form (7.9)

in which $\alpha = 4K'\Sigma_0/\delta$, where Σ_0 is the area of the wall.*

Another possible consistent motion of the spins in a wall which is not associated with a change in $\mathcal{H}_{ex}$ and $\mathcal{H}_{an}$ involves simultaneous rotation of the spins about the easy axis z (the angles θ_j between the spins and the z axis, and also the angles between the spins themselves, are preserved). However, a component M_y of the local magnetization normal to the plane of the wall will thereby appear, increasing the magnetostatic energy. If the spins do not lie in the plane of the wall, a demagnetizing field $H_y = -4\pi M_y$ appears, tending to restore the spins to the XZ plane. The corresponding energy density can approximately be presented in the form

$$\mathcal{H}_{mst} = -\tfrac{1}{2} M_y H_y = 2\pi M_y^2 . \tag{7.12}$$

When the energies $\mathcal{H}'_{an}$ and $\mathcal{H}_{mst}$ are taken into account, the directions of the spins in a wall are fully stabilized with respect to small perturbations: $\mathcal{H}'_{an}$ hinders the deviation of M from the plane Zy, and $\mathcal{H}_{mst}$—from the plane ZX. In the equilibrium state, $M_X = M_y = 0$ and $M_Z = M_0$. Therefore the full energy density

$$\mathcal{H}(r) = \mathcal{H}_{ex} + \mathcal{H}_{an} + \mathcal{H}'_{an} + \mathcal{H}_{mst}$$

expressed in terms of the components M_X, M_y and M_Z can be expanded in powers of M_X and M_y near this state. The linear terms vanish on account of the stability of the ground state, while the quadratic terms can be brought to the form

$$\mathcal{H}_2(r) = \frac{J_0 a^2}{M_0^2} [(\nabla M_x)^2 + (\nabla M_y)^2] + \frac{K}{M_0^2} \cos 2\theta (M_X^2 + M_y^2) + \frac{K'}{M_0^2} M_X^2 + 2\pi M_y^2 . \tag{7.13}$$

Terms of higher order can be neglected in energy spectrum computations. They are responsible for relaxation processes in the spin system.

The spectrum can be computed by writing down and solving the classical equations of motion for the local magnetization (Winter, 1964). It is, however, of interest to carry out a quantum-mechanical analysis, which is especially important for the investigation of kinetic phenomena associated with domain walls. It turns out that the full energy of small vibrations of the magnetic moments (near their equilibrium

---

* We have used the relation $\int \sin^2 \theta \, dV = 2\Sigma_0 \delta$.

position) $\mathcal{H}_2 = \int \mathcal{H}_2(r)\,dV$, as in the case of a homogeneously magnetized spec-imen, can be reduced to a sum of the energies of "elementary excitations," magnons, of two types,

$$\mathcal{H}_2 = \sum_\kappa \varepsilon_{1\kappa} n_{1\kappa} + \sum_k \varepsilon_{2k} n_{2k} + \Delta\mathcal{H}_0. \tag{7.14}$$

Here $\varepsilon_{1\kappa}$ and ε_{2k} are the excitation energies of the first and second species, respectively ($\Delta\mathcal{H}_0$ is the energy of the zero-point vibrations). It is characteristic that $\varepsilon_{1\kappa} \equiv \varepsilon_1(\kappa)$ depends on the two-dimensional wave vector $\kappa \equiv (\kappa_x, \kappa_z)$ lying in the plane of the wall, while $\varepsilon_{2k} = \varepsilon_2(k)$ depends on the ordinary three-dimensional wave vector $k \equiv (k_x, k_y, k_z)$. The explicit expressions for these energies are the following:

$$\varepsilon_{1\kappa} = \frac{2\mu_e}{M_0}\{[J_0(a\kappa)^2 + K'][J_0(a\kappa)^2 + 2\pi M_0^2]\}^{1/2}, \tag{7.15}$$

$$\varepsilon_{2k} = \frac{2\mu_e}{M_0}\{[J_0(ak)^2 + K + K'][J_0(ak)^2 + K + 2\pi M_0^2]\}^{1/2} \tag{7.16}$$

$$(\mu_e = |\gamma_e|\hbar).$$

The number of magnons of each species (with wave vector κ or k) can be expressed in the usual form through the appropriate particle creation and annihilation oper-ators: $n_{1\kappa} = a_{1\kappa}^+ a_{1\kappa}$ or $n_{2k} = a_{2k}^+ a_{2k}$.

For the computation of the various correlation functions, we require, in addition to the energies (7.15) and (7.16), also the transformation which expresses the trans-verse components of the local magnetization M_X and M_y (in the second quantization representation) in terms of the operators $a_{1\kappa}^+$, $a_{1\kappa}$ and a_{2k}^+, a_{2k}. It has the form

$$M^+(r) = M_X + iM_y = \left(\frac{2\mu_e M_0}{\Sigma_0}\right)^{1/2} \sum_\kappa e^{-i\kappa r}\sin\theta(y)(U_{1\kappa}a_{1\kappa}^+ + V_{1\kappa}a_{1,-\kappa}) +$$

$$+ \left(\frac{2\mu_e M_0}{\Sigma_0}\right)^{1/2} \sum_k e^{-ikr}[\cos\theta(y) - ik_y\delta](U_{2k}a_{2k}^+ + V_{2k}a_{2,-k}). \tag{7.17}$$

(The analogous relation for $M^-(r) = M_X - iM_y$ can be obtained from (7.17) by the substitution $a^+ \leftrightarrow a$ with a subsequent change in the signs of κ and k). The coefficients U and V are defined by the following expressions:

$$
\begin{aligned}
U_{1\kappa} = \\
V_{1\kappa} =
\end{aligned}
\frac{1}{2\sqrt{\delta}}\left\{\frac{\dfrac{2\mu_e J_0}{M_0}(a\kappa)^2 + 2\pi\mu_e M_0 + \dfrac{\mu_e K'}{M_0}}{\varepsilon_{1\kappa}} \pm 1\right\}^{1/2},
$$

$$U_{2k} = \frac{1}{\sqrt{2\left[L(1 + k_y^2\delta^2) - 2\delta\right]}} \times$$
$$V_{2k} =$$

$$\times \left\{ \frac{\dfrac{2\mu_e J_0}{M_0}(ak)^2 + \dfrac{2\mu_e(K + K'/2)}{M_0} + 2\pi\mu_e M_0}{\varepsilon_{2k}} \pm 1 \right\}^{1/2}, \qquad (7.19)$$

where L is the linear dimension of the specimen in the direction of the y axis; in both equations, the plus and minus signs refer to U and V, respectively.

The longitudinal (in the local coordinate system) component of M can also be expressed through the operators a and a^+ with the help of the relation

$$M_z(r) = M_0 - M^+M^-/2M_0. \qquad (7.20)$$

It is not difficult to prove that the transformation (7.17), together with the analogous transformation for M^-, actually reduces the energy $\mathcal{H}_2 = \int \mathcal{H}_2(r)\, dV$ to the "diagonal" form (7.14), provided the operators a and a^+ satisfy the Bose commutation relations

$$a_{1\kappa}a_{1\kappa'}^+ - a_{1\kappa'}^+a_{1\kappa} = \delta_{\kappa\kappa'}, \text{ etc.}$$

Consequently, the mean numbers of magnons $\langle n_{1\kappa}\rangle$ and $\langle n_{2k}\rangle$ are again determined by the Bose distribution function (2.17) with the respective energies $\varepsilon_{1\kappa}$ and ε_{2k}.

From the classical point of view, the spins in domain walls can exhibit two vibration modes. For one branch of vibrations, the eigenfrequency $\omega_{1\kappa} = \varepsilon_{1\kappa}/\hbar$ depends on the two-dimensional wave vector κ lying in the plane of the wall, and the lowest frequency of these vibrations (for $\kappa = 0$)

$$\Delta \equiv \omega_{10} = |\gamma_e|\sqrt{8\pi K'} \qquad (7.21)$$

corresponds to homogeneous elastic displacements (translations) of the wall as a whole in the direction of the y axis. As is seen from (7.17), the amplitudes of these vibrations are proportional to $\sin\theta(y) = 1/\cosh(y/\delta)$. Consequently, they are maximum in the center of the wall (for $y = 0$) and vanish as $e^{-|y|/\delta}$ outside the wall (for $|y| \gg \delta$). Excitations of this type are thus the specific excitations of electron spins which are characteristic of domain walls.

Vibrations of the second type constitute a modified form of the ordinary spin waves in a single-domain ferromagnet for the transition region between domains. The lowest frequency for this branch is obtained from (7.16) with $k = 0$,

$$\omega_{20} = \varepsilon_{20}/\hbar = |\gamma_e|\sqrt{\frac{2(K + K')}{M_0}\left(4\pi M_0 + \frac{2K}{M_0}\right)}. \qquad (7.22)$$

Usually $K \gg K'$, so that $\omega_{20} \gg \omega_{10}$. Moreover, according to (7.17), the absolute magnitude of the amplitude of these vibrations is proportional to $\left[\left(\tanh \dfrac{y}{\delta}\right)^2 + k_y^2 \delta^2\right]^{1/2}$ and, consequently, it has a minimum inside the wall (vanishing at the center of the wall for $k_y \to 0$). For this reason, vibrations of the first ("in-wall") type play the principal role in NMR phenomena involving nuclei inside domain walls. In what follows, we shall take only these into consideration.

"In-wall" excitations can have a strong effect on the thermodynamic properties of spins in domain walls. In particular, they lead to different temperature dependences of the local magnetization (and, consequently, of the NMR frequency $\omega_n(T) = \gamma_n A_0 \langle M_z \rangle$) for domains and for walls. Because the wave vector κ is two-dimensional, the first term in the expansion of $\Delta M(T)$ in terms of temperature is linear, and not a power of $\frac{3}{2}$. Indeed, calculating only the contribution of the "in-wall" excitations to $\Delta M(T)$, we have, according to (7.20) and (7.17),

$$\langle M_z \rangle = M_0 - (\mu_e/\Sigma_0) \sin^2 \theta \sum_{\kappa} \left[V_{1\kappa}^2 + (U_{1\kappa}^2 + V_{1\kappa}^2) \langle n_{1\kappa} \rangle\right]. \tag{7.23}$$

The first term in the sum represents a "cancellation" of spins, due to the zero-point vibrations at $T = 0° \text{K}$. To orders of magnitude, it is equal (at the center of the wall) to

$$\frac{(\Delta M)_0}{M_0} \sim \left(\frac{K}{J_0}\right)^{1/2} \frac{\mu_B M_0}{J_0 a^3}. \tag{7.24}$$

This "cancellation" of spins can provide still another reason for the different NMR frequencies of a wall and a domain for $T \to 0° \text{K}$.

The decrease of $\langle M_z \rangle$ caused by "in-wall" thermal excitations, which is produced by the term with $\langle n_{1\kappa} \rangle$ in (7.23), takes the form (for $\kappa T \gg 2\pi \mu_e M_0$ and $\mu_e K'/M_0$)

$$\frac{\Delta M(T)}{M_0} = \left(\frac{K}{J_0}\right)^{1/2} \frac{\kappa T}{16\pi J_0 a^3} \ln\left[\frac{\kappa T}{\mu_e \left(\dfrac{K'}{M_0} + 2\pi M_0\right)}\right] \sin^2 \theta(y). \tag{7.25}$$

Note that the difference in the NMR frequencies for walls and domains observed in $CrBr_3$ (see Figure 15) indeed varies linearly with temperature.

The decrease of local magnetization caused by "in-wall" excitations is proportional to $\sin^2 \theta(y)$ and, consequently, it is not uniform across the wall thickness; it thus naturally leads not only to a shift in the NMR line, but can also affect its width in some cases.

3. Line Width and Relaxation

For nuclei in domain walls both the resonance frequencies and the amplification coefficient η depend on the direction of the local magnetization. Moreover, as η is, in general, a complex quantity, so that the NMR signal for each nuclear spin is a superposition of absorption and dispersion curves (see Chapter III, Sec. 1), the resultant absorption curve due to nuclei in a domain wall usually has a very complicated shape which is generally asymmetric and depends on the intensity of the r.f. field. The line may also have a fine structure on account of quadrupole splitting.*

Here we shall consider only the specific mechanism of NMR line broadening and relaxation which is associated with the existence of "in-wall" spin excitations. As in the single-domain case, the indirect (Suhl–Nakamura) interaction between nuclear spins via spin waves is one of the basic NMR line broadening mechanisms for high concentrations of magnetic nuclei. Excitations of the "in-wall" type, which have a smaller energy gap, again play the main role.

The indirect interaction between nuclear spins in domain walls is calculated as in Chapter IV, Sec. 1, using (5.72) and (7.17) from which the HFI energy is expressed in terms of the creation and annihilation operators of electron spin excitations in the wall ($a_{1\kappa}^{+}$ and $a_{1\kappa}$). Referring the reader for details to the work of Winter (1961), we shall only note that the effective interaction radius (b_0) of nuclear spins in a domain wall in the directions Z and X is larger than inside a domain. In the case when $\omega_n < \Delta$ this radius is obtained from the corresponding magnitude for a single-domain specimen (4.15) by the simple substitution $K \to K'$ (for $H = 0$) and consequently has the form $b_0 = a \sqrt{\dfrac{J_0}{K'}}$. The second moment of the NMR line for a wall is accordingly $(K/K')^{1/2}$ times larger than that calculated by Suhl (1958) for a single domain ferromagnet (see expression (4.19)).

Expressing $\delta H^{\pm} = - A_0 M^{\pm}$ and $\delta H^{Z} = - A_0(M_Z - \langle M_Z \rangle)$ in terms of the operators $a_{1\kappa}^{+}$ and $a_{1\kappa}$ with the help of (7.17) and (7.20), we can then calculate from (5.12)–(5.15) the rates of longitudinal and transverse relaxation due to the scattering of "in-wall" excitations by nuclear spins. From the classical point of view, the hyperfine field fluctuations which determine these processes are caused by the thermal motion of the domain walls. Such processes in a number of cases are quite significant for the longitudinal relaxation of nuclear spins in domain walls.

Since $M^{\pm}$ is only determined by expressions of the form (7.17) up to terms linear in the operators a^{+} and a, we can consider in this approximation only one-phonon

---

* A detailed discussion of the problem of the resultant NMR spectrum in domain walls (taking the dipole and the quadrupole interactions and the nonhomogeneity of the coefficient η into account) for ferromagnets of different symmetries and domain structures is given by Murray and Marshall (1965).

processes of spin-lattice relaxation. In terms of free magnons, they are allowed by the law of energy conservation only if $\omega_n > \Delta$. Since the electron resonance frequency of a domain wall Δ may be considerably lower than the FMR frequency of a single-domain specimen, this condition is not unlikely. Then

$$\frac{1}{T_1} \simeq \tfrac{1}{8} \sin^2 \theta(y)\, \omega_n \sqrt{\frac{K}{J_0}} \left(\frac{\kappa T}{J_0 a^3} \right). \tag{7.26}$$

If $\omega_n < \Delta$, one-magnon processes are possible only when the damping of "in-wall" excitations is taken into account. In the same approximations as in the derivation of the analogous expression (5.56) for a single-domain specimen, we have

$$\frac{1}{T_1} \simeq \Gamma \frac{\sin^2 \theta}{16\pi} \left(\frac{\omega_n}{\Delta} \right)^2 \left(\frac{K}{J_0} \right)^{1/2} \frac{\kappa T}{J_0 a^3}, \tag{7.27}$$

where Γ is some effective viscosity coefficient of the wall at the frequency $\omega = \omega_n$.

Quantitative estimates of $1/T_1$ are difficult to obtain because of the uncertainty in the parameters Δ and Γ. In metals, damping is basically associated with eddy currents and is therefore sensitive to the grain size in the specimen. Using reasonable values of the parameters, Winter (1961) obtained for Fe and Ni at room temperature (for the center of the wall, where $\sin^2 \theta(y) = 1$) $T_1 \sim 10^{-3} - 10^{-4}$ sec, which is comparable with experimental values. In addition to its linear dependence on temperature, $1/T_1$ also depends on temperature through Γ, K, and Δ.

Le Dang Khoi (1965 and 1966) obtained confirmation of the importance of this nuclear spin relaxation mechanism in domain walls by varying the parameters which enter the analogs of (7.27) for iron and yttrium–gadolinium ferrite–garnets.

For a high concentration of magnetic nuclei (e.g., in cobalt), the Suhl–Nakamura interaction between nuclear spins has a strong effect on the observed spin-lattice relaxation time. Portis and Gossard (1960) showed that in NMR involving nuclei in domain walls, nuclear spins in domains (outside the walls) are hardly excited (because of the difference in the resonance frequency and the considerably smaller amplification coefficient). Therefore, the Suhl–Nakamura interaction, as it were, produces diffusion of the nuclear spin excitations induced by the r.f. field from walls into domains. As a result, the magnetic energy of the nuclei does not only relax to the lattice inside the wall (by means of the fluctuation mechanism considered in this section, or via the interaction with collective electrons considered in Chapter V, Sec. 6), but is also transported from the wall by spin diffusion. This must be borne in mind when computing the true spin-lattice relaxation time τ_l, since the observed time T_1 may considerably differ from τ_l both in magnitude and in its temperature dependence.

CHAPTER VIII

The Chemical Bond and Nuclear Magnetic Resonance of Nonmagnetic Ions

1. Phenomenological Description of the HFI of the Nuclei of Non-magnetic Ions. Spin Density

The local magnetic fields are often measured at the nuclei of nonmagnetic ions. The nature of these fields is generally the same as the nature of the local fields at the nuclei of magnetic ions, i.e., they are produced by dipole–dipole and hyperfine interactions.

At the nuclei of nonmagnetic ions, as distinct from magnetic ions, the hyperfine and the dipole fields may be of the same order of magnitude, and in a number of cases H_{dip} may even predominate. We have mentioned before (Chapter II, Sec. 1) that the magnitude and the direction of the dipole field at a fixed temperature are determined by the magnetic structure and the interatomic distances, whereas the temperature dependence of H_{dip} is determined by the temperature dependence of the sublattice magnetizations, if the change in interatomic distances is ignored. Experimental investigations of dipole fields at the nuclei of nonmagnetic ions may thus also be used for the determination of magnetic structures, Néel points, and the temperature dependence of the sublattice magnetizations. The most characteristic results in this respect are provided by the measurement of the NMR of protons, first carried out in $CuCl_2 \cdot 2H_2O$ (Poulis et al., 1952, 1953, 1958), and later in $CoCl_2 \cdot 2H_2O$ and $FeCl_2 \cdot 2H_2O$ (Narath et al., 1964; Narath, 1965a; Spence et al., 1964).

These authors showed how the magnetic structure of the antiferromagnets can be determined from the angular dependence of the NMR spectra; the Néel temperatures were also determined in some of the materials. The local fields at the protons in these

crystals, which are produced almost entirely by dipole–dipole interactions, attain a few kilooersted in the magnetically ordered phase. Measurements can thus be made without the application of an external magnetic field.

We shall consider in more detail those cases in which the largest contribution to the local field at the nuclei of nonmagnetic ions is from the hyperfine interaction. HFI effects were first discovered for fluorine nuclei in MnF_2 by measurements in a paramagnetic phase (Shulman and Jaccarino, 1956, 1957). Subsequently they were studied in a large number of paramagnetic and magnetically ordered crystals at the nuclei of chlorine ions (Narath, 1965b), bromine ions (Gossard et al., 1964), iodine ions (Burgiel et al., 1961), and oxygen ions (O'Reily and Tsang, 1964) which surround the metallic ions and participate in the chemical bond (the so-called ligands).

Fairly recently, hyperfine fields were observed by means of NMR at nuclei of nonmagnetic metal cations, e.g., gallium nuclei in the ferrite–garnet $Y_3Fe_{5-x}Ga_xO_{12}$ (Streever and Uriano, 1964, 1965) and Tl and Rb nuclei in paramagnetic $TlMnF_3$ (Petrov and Smolenskii, 1965), $RbMnF_3$ (Payne et al., 1965; Petrov et al., 1965; Wolker and Stevenson, 1966), and $RbCoF_3$ (Petrov and Kudryashov, 1966; Petrov and Nedlin, 1967). Analogous fields were observed by means of the Mössbauer effect at tin nuclei in the tin-doped ferrite–garnet $Y_3Fe_5O_{12}$ (Goldanskii et al., 1965; Belov and Lyubutin, 1965).

The appearance of hyperfine fields at the nuclei of nonmagnetic ions shows quite unambiguously that the electron shells of these ions are uncompensated and that there exists spin polarization near these nuclei.

Measurements of these fields allow direct observation of the perturbation of closed electron shells of nonmagnetic ions in magnetic crystals. The information about the character and magnitude of this perturbation is of fundamental importance, at least for two problems in the physics of magnetic crystals.

The first problem concerns the role of nonmagnetic atoms in the superexchange interaction responsible for magnetic ordering. The concept of a superexchange interaction, first advanced by Kramers (1934) and subsequently developed by many authors (Review of Anderson, 1963), assumes a long-range exchange interaction via intermediate atoms.* This idea proved very fruitful for our understanding of the nature of magnetic ordering in dielectric crystals, where other exchange mechanisms, e.g., those caused by the conduction electrons, are absent. However, a direct verification of the theory is difficult because of the lack of experimental data about the character and the details of the interaction between the electron shells of magnetic and nonmagnetic ions.

---

* In the Soviet literature, the exchange interaction via intermediate ions is often called *indirect exchange*. Anderson, analyzing various exchange mechanisms, uses the term "indirect exchange" for some of them, and the term "superexchange" for others. Roughly speaking, indirect exchange involves excited ligand states, whereas superexchange is associated with the change in the form of the paramagnetic ion wave function caused by admixture of the ligand function.

NMR measurements fill this gap to a considerable extent. The investigation of the HFI of nuclei of nonmagnetic ions provides quantitative data on the mixing of wave functions of magnetic and nonmagnetic ions. This information can be very useful for the substantiation and verification of the theory of the superexchange interaction, leading to magnetic ordering in dielectrics.

The second problem concerns the spatial distribution of spin density in crystals. Measurements of fields and spin densities at various nuclear species in the lattice permit determining the topology of spin and electron density in a crystal. An example of such an investigation will be given in Secs. 3 and 4, where some work will be described on the study of spatial distribution of spin density in paramagnetic fluorides.

The general phenomenological description of the hyperfine interaction of nuclei of nonmagnetic ions (or the hyperfine fields) can be formulated either in terms of a set of HFI constants $A_{\alpha\alpha'}$, or in terms of "spin density transport." When we wish to compare the hyperfine interactions of nonmagnetic ions in different crystals and identify the contribution from the chemical bond and the electron shell overlap, it is convenient to use the concept of spin density. We shall now give a definition of spin density in terms of the constants $A_{\alpha\alpha'}$, and in the next sections we shall show how the spin density is related to covalency and overlap and how it can be determined from experiment.

As an example, consider the octahedral complex $[MF_6]^{4-}$ (Figure 16), where M is a paramagnetic ion of the $3d$-group. The octahedral coordination is the one most characteristic for atoms of the transitional $3d$-group. Such octahedra (regular or somewhat distorted) are observed in crystals of widely differing structures and symmetries. As ligands it is advisable to take ions with the simplest electron structure, e.g., fluorine ions.

The cubic symmetry of the surroundings of the ion M enables the local coordinate axes σ, π and π' to be chosen at the fluorine positions as shown in Figure 16. The σ axis pointing to the central ion coincides with the tetragonal axis of the complex. All the nuclei in the complex are equivalent but, to be specific, we shall consider the nucleus in position 3 (the axes π, π' and $-\sigma$ for this nucleus coincide with the X, Y and Z axes of the general coordinate system). The results obtained for this particular case can be easily generalized to any fluorine nucleus in the octahedron by a simple rotation of the coordinate axes.

The HFI of any nucleus with a central paramagnetic ion can be written in the local coordinate system in the form

$$\mathscr{H} = A_{\parallel} I^\sigma S^\sigma + A_{\perp}(I^\pi S^\pi + I^{\pi'} S^{\pi'}). \tag{8.1}$$

Let the electron and the nuclear spin quantization axes coincide, both pointing along the Z axis, parallel to the external field. Then only the Z-component of the electron spin contributes to the mean value of the hyperfine field at the nucleus;

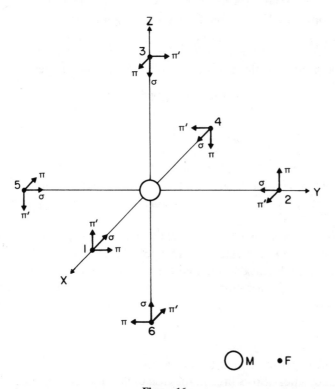

Figure 16

The octahedral complex $[MF_6]^{4-}$ with the paramagnetic atom at the center, surrounded by six fluorine atoms. The axes of the local coordinate system (σ, π, and π') shown for each fluorine atom are chosen so that the axis σ points to the central atom.

in this case,

$$H^z_{HFI} = - (\gamma_n \hbar)^{-1} \left[A_p (3 \cos^2 \theta_{\sigma z} - 1) + A_s \right] \bar{S}^z . \qquad (8.2)$$

Here $A_s = \frac{1}{3}(A_\parallel + 2A_\perp)$ and $A_p = \frac{1}{3}(A_\parallel - A_\perp)$ are constants describing the isotropic and the anisotropic HFI respectively, $\theta_{\sigma z}$ is the angle between the σ and Z axes, and $\bar{S}^z$ is the quantum-mechanical mean value of the Z-projection of the spin in the ground state of the paramagnetic ion. For paramagnetic ions with a locked orbital momentum or with $L = 0$, $\bar{S}^z$ is simply the maximal projection of the spin S.

In order to introduce the concept of spin density, we compare the hyperfine field (8.2) with the local fields produced at the nucleus by the individual electrons in the shells of an isolated fluorine ion.

For example, one electron in the $2s$ shell with spin projection $\bar{S}^z = \frac{1}{2}$ (spin "up") creates a field

$$H^z_{2s} = - (\gamma_n \hbar)^{-1} a_{2s} \tfrac{1}{2} , \qquad (8.3)$$

and an electron with $\bar{S}^z = \frac{1}{2}$ in the $2p_\alpha$ shell creates a field

$$H^z_{2p\alpha} = -(\gamma_n\hbar)^{-1} a_{2p} (3\cos^2\theta_{\alpha z} - 1)\tfrac{1}{2} \tag{8.4}$$

($\alpha = \sigma$, π, or π' are the three local axes (see Figure 16)). Here

$$a_{2s} = \tfrac{16}{3}\pi\mu_B\gamma_n\hbar \,|\psi_{2s}(0)|^2 = 1.503 \text{ cm}^{-1},$$

$$a_{2p} = \tfrac{4}{5}\mu_B\gamma_n\hbar \left(\frac{1}{r^3}\right)_m = 0.0429 \text{ cm}^{-1} \tag{8.5}$$

(Shulman and Sugano, 1963; Abragam, 1963), $(1/r^3)_m$ is the quantum-mechanical mean value of $(1/r^3)$ for the $2p$ shell, and μ_B is the Bohr magneton.

The fields set up by an electron with a spin "down" ($\bar{S}^z = -\frac{1}{2}$) have an opposite sign.

In an isolated fluorine ion, the total field from all the electron spins in every shell vanishes, since the fields created by the "up" (↑) and "down" (↓) spins mutually cancel. In a crystal, because of the interaction with the electron shells of the para-magnetic ions, spin polarization occurs in the fluorine ion shells, and the electron wave functions with spins "up" and "down" will be different at a given point r near the fluorine ion nucleus,

$$|\psi^\uparrow(r)|^2 \neq |\psi^\downarrow(r)|^2. \tag{8.5a}$$

As a result, the hyperfine field at the fluorine ion nucleus is no longer zero.

A direct calculation shows that, in the case of fluorine ions, the main contribution to the hyperfine field comes from the polarization of the $2s$ and $2p$ shells. It can be assumed (this is proved by further calculations) that the contributions from the various shells will be proportional to the maximal fields (8.3) and (8.4). In other words, we shall assume that the total hyperfine field at the fluorine ion nucleus is

$$H^z_{\text{HFI}} = \sum_\alpha f_\alpha H^z_{2p_\alpha} + f_s H^z_{2s}, \tag{8.6}$$

where the coefficients f_α are the parameters which characterize the degree of polariza-tion of the corresponding shells. We shall call these f_α the spin densities.*

Our next task is to relate these parameters to the phenomenological HFI constants A_p and A_s and to the quantum-mechanical parameters characterizing the chemical bond, i.e., the covalency and the overlap parameters.

The first task is quickly accomplished by a direct comparison of (8.1) and (8.6),

---

* For the s shell, where the hyperfine field is determined by the Fermi contact interaction,

$$f_s = |\psi^\uparrow_{2s}(0)|^2 - |\psi^\downarrow_{2s}(0)|^2.$$

Consequently, f_s is indeed a local characteristic of the spin density at the nucleus ($r = 0$). In general f_α is also an integral characteristic of the spin polarization of the corresponding shell in some region with dimensions of the order of the $2p$-orbital radius.

taking (8.3) and (8.4) into account. Using the relation $f_\pi = f_{\pi'}$, which follows from the tetragonal symmetry of the fluorine positions, we readily find

$$A_s = f_s a_{2s} \cdot \frac{1}{2\bar{S}^z},$$
(8.7)

$$A_p = (f_\sigma - f_\pi) a_{2p} \frac{1}{2\bar{S}^z}.$$
(8.8)

To solve the second part of the problem, we require the wave function of the complex.

2. Calculation of Spin Polarization by the Method of Molecular Orbitals

Two methods are commonly employed for a quantum-mechanical analysis of the effects which produce spin polarization in ligand shells: the method of configurational interaction and the method of molecular orbitals.

In the first method, a pure ionic state is chosen as the ground state: M^{++} for the metal and F^- for fluorine. In the ground state, the electron shells of fluorine are fully compensated. An excited state is considered corresponding, e.g., to a configuration in which an electron is transferred from a fluorine shell to a metal shell, i.e., the configuration M^+—F. In this configuration there is an unpaired electron spin in the fluorine shell, which sets up an effective field at the nucleus.

In the second method the change in the ion wave functions is taken into consideration, which leads already in the ground state to spin polarization in fluorine.

Calculated values of spin density close enough to those observed experimentally (Rimmer, 1965; Hubbard et al., 1966) have been obtained by the method of configurational interaction. However, the method of molecular orbitals is more widespread and we shall therefore examine it in more detail. Some comments on the difference in the interpretation of the experimental data by the two methods will be found in Sec. 5.

The method of molecular orbitals constitutes one of the variational methods for the solution of the many-electron problem in a one-electron approximation. Atomic or ionic wave functions are usually used as trial functions (although this is not obligatory). This method was first used for the investigation of the electron structure of molecules and subsequently Van Vleck (1935) applied it to complex compounds. The method of molecular orbitals is now frequently used also for finding the wave functions in nominally ionic crystals.

The basic idea behind the application of this method (in particular, the LCAO approximation—linear combinations of atomic orbitals) to ionic crystals is the following (Van Vleck, 1935; Ballhausen, 1964). A part of the crystal (a complex) is

selected, containing a central ion and its closest surroundings. The symmetry of this complex must correspond to the local symmetry of the crystal.

It is assumed that the coordinate part of the wave function of the many-electron state of the complex can be represented in the form of a determinant of one-electron functions, molecular orbitals (MO). In the construction of a MO, the electrons are assumed not to interact among themselves. The structural unit for MO construction is the entire complex, and not the individual atoms. An MO must naturally belong to some irreducible representation of the symmetry group of the complex.

In the LCAO approximation, the molecular orbitals have the form

$$\Psi_{MO} = \Psi(\Gamma) + \sum_i c_i \Psi_i, \tag{8.9}$$

where $\Psi(\Gamma)$ is a one-electron wave function of the central ion which belongs to the irreducible representation Γ. The sum $\Sigma c_i \Psi_i$ is a linear combination of one-electron wave functions of the surrounding atoms which belongs to the same representation.

The coefficients c_i in principle can be found either by the variational method or from experiments, for example, by NMR.

As an example, we shall consider the negatively charged octahedral complex $[\mathrm{NiF_6}]^{4-}$ (see Figure 16). We shall construct the MO in the form of a combination of one-electron wave functions of the nickel shell and the $2p$- and $2s$-functions of the fluorine. It is well known (Ballhausen, 1964) that one-electron functions of the central atom (nickel) in a field of octahedral symmetry belong to the two-dimensional and the three-dimensional representations e_g and t_{2g}, respectively.* Therefore the MO incorporating the $3d$-functions of nickel and the $2p$- and $2s$-functions of fluorine should also belong to e_g- or t_{2g}-representations.

An explicit form of the relevant MO for the complex $[\mathrm{NiF_6}]^{4-}$ is given by Shulman and Sugano (1963):

$$e_g\text{-orbitals} \quad \begin{cases} \Psi_e^a = N_e^{-1/2}(\phi_e - \lambda_s \chi_s - \lambda_\sigma \chi_\sigma), \\[6pt] \Psi_{es}^b = N_e'^{-1/2}(\chi_s + \gamma_s \phi_e), \\[6pt] \Psi_{e\sigma}^b = N_e''^{-1/2}(\chi_\sigma + \gamma_\sigma \phi_e), \end{cases} \tag{8.10}$$

$$t_{2g}\text{-orbitals} \quad \begin{cases} \Psi_t^a = N_t^{-1/2}(\phi_t - \lambda_\pi \chi_\pi), \\[6pt] \Psi_t^b = N_t'^{-1/2}(\chi_\pi + \gamma_\pi \phi_t). \end{cases} \tag{8.11}$$

In these expressions the following notation is used: ϕ_e denotes one of the two degenerate one-electron orbitals of nickel, transforming as $3Z^2 - r^2$ and $X^2 - Y^2$; ϕ_t is one of the three degenerate orbitals transforming as XY, YZ, XZ (in order not

---

* We follow the notation of Mulliken (1933), whereby small letters are used to describe one-electron states, and capital letters are used for many-electron states.

to complicate the notation, we shall omit the corresponding indices for ϕ_e and ϕ_t); finally, χ_s, χ_σ, and χ_π are the molecular orbitals of fluorine, which are defined as linear combinations of the fluorine atomic functions:

$$\chi_{s(3Z^2-r^2)} = \frac{1}{\sqrt{12}}(2s_3 + 2s_6 - s_1 - s_2 - s_4 - s_5),$$

$$\chi_{s(X^2-Y^2)} = \tfrac{1}{2}(s_1 + s_4 - s_2 - s_5), \qquad (8.12)$$

$$\chi_{\sigma(3Z^2-r^2)} = \frac{1}{\sqrt{12}}(2\sigma_3 + 2\sigma_6 - \sigma_1 - \sigma_2 - \sigma_4 - \sigma_5),$$

where s_i and σ_i are the 2s- and 2p-functions of the i-th fluorine. Further,

$$\chi_{\pi(XY)} = \tfrac{1}{2}(\pi_1 + \pi'_4 + \pi'_2 + \pi_5),$$

$$\chi_{\pi(XZ)} = \tfrac{1}{2}(\pi_3 + \pi'_1 + \pi_4 + \pi'_6), \qquad (8.13)$$

$$\chi_{\pi(YZ)} = \tfrac{1}{2}(\pi_2 + \pi'_3 + \pi_6 + \pi'_5).$$

Here π_i and π'_i are respectively the $2p_\pi$- and $2p_{\pi'}$-functions of fluorine; the direction of the axes is the same as in Figure 16.*

In (8.10) and (8.11) the following normalization coefficients are introduced:

$$N_e = 1 - 2\lambda_s S_s - 2\lambda_\sigma S_\sigma + \lambda_s^2 + \lambda_\sigma^2 + 2\lambda_s S_\sigma,$$

$$N'_e = 1 + 2\gamma_s S_s + \gamma_s^2,$$

$$N''_e = 1 + 2\gamma_\sigma S_\sigma + \gamma_\sigma^2, \qquad (8.14)$$

$$N_t = 1 - 2\lambda_\pi S_\pi + \lambda_\pi^2,$$

$$N'_t = 1 + 2\gamma_\pi S_\pi + \gamma_\pi^2.$$

The normalization coefficients incorporate the so-called group overlap integrals

$$S_s = \int \phi_e \chi_s d\tau, \qquad S_\sigma = \int \phi_e \chi_\sigma d\tau, \qquad S_\pi = \int \phi_t \chi_\pi d\tau, \qquad (8.15)$$

characterizing the overlap between the wave functions of nickel and the molecular orbitals of fluorine. The latter can be expressed in terms of the atomic overlap integrals (Ballhausen, 1964) and the atomic wave functions. In our case of cubic

---

* The indices σ and π denote the absolute value of the quantum number m_e of the projection of the electron orbital momentum l on the axis joining the two atoms: σ corresponds to $m_e = 0$, and π to $m_e = 1$ (Penney et al., 1938). The same indices are also used to denote the type of chemical bond. The σ-bond is made up of wave functions stretched along the bond line, and the π-bond is made up of wave functions stretched in a direction perpendicular to the bond.

symmetry,

$$S_s = \sqrt{3} \int \phi_{e(3Z^2-r^2)} s_3 d\tau,$$

(8.16)

$$S_\pi = 2 \int \phi_{t(XY)} \cdot \pi_1 d\tau, \text{ etc.}$$

Indices a and b in (8.10) and (8.11) denote antibonding and bonding orbitals. Antibonding orbitals are defined as orbitals whose energy increases when the atoms are brought closer to each other, while for bonding orbitals the energy decreases when the interatomic distance is reduced. It is because of the presence of bonding orbitals that stable compounds are formed.

The orbitals e_g and t_{2g} are automatically orthogonal, since they belong to different representations. However, for bonding and antibonding orbitals of the same symmetry to be orthogonal, the following conditions must be fulfilled:

$$\lambda_s = \gamma_s + S_s, \quad \lambda_\sigma = \gamma_\sigma + S_\sigma, \quad \lambda_\pi = S_\pi + \gamma_\pi \quad (8.17)$$

(in writing these conditions, terms quadratic in γ and S were dropped).

It now remains to explain the meaning of the quantities γ_s, γ_σ, and γ_π. These quantities characterize the degree of covalency of the chemical bond. If $\gamma = 0$, the corresponding bond is purely ionic, and if $\gamma = 1$, it is purely covalent. Note that these covalency parameters determine the probability of finding an electron on one of the atoms and constitute a characteristic of a single electron participating in the chemical bond, but not of the ion as a whole, i.e., one should not use γ to make inference about the affinity of the ion as a whole to form covalent bonds.

To solve the problem completely, it is now necessary to find the energy eigenvalues corresponding to each MO. We shall not consider the details of the calculations which lead up to the level diagram shown in Figure 17. The only significant point is that the bonding orbitals lie below the free atom levels.

Let us consider the sequence in which the electron orbitals are filled. Since we are not interested in all the MO of the complex but only in those which include the $3d$-orbitals of the central atom, we shall examine the placement of the d-electrons only, for various configurations of the $3d$-shell. If the shell contains no electrons (the d^0-configuration), the bonding orbitals are filled (with electrons supplied by fluorine), while the antibonding orbitals are empty. In filling the molecular orbitals, one can be guided by the same considerations as in filling one-electron orbitals in the theory of the crystal field (Ballhausen, 1964). For configurations from $3d^0$ to $3d^3$, the degenerate t_{2g}^a-orbitals are being filled by three electrons with parallel spins. For the $3d^4$ configuration, the fourth electron can be placed either in a t_{2g}^a-orbital or in a e_{2g}^a-orbital. In the former case, the spin of the fourth electron will be opposite to the

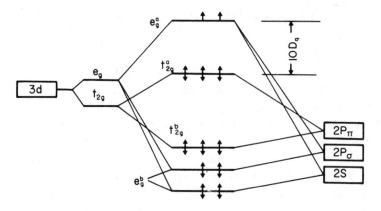

Figure 17

A diagram of molecular orbital levels.

spins of the first three, and in the latter case the spins will be parallel. The particular case observed in practice depends on the relation of the intra-atomic electron inter-action and the energy gap $\Delta = E(e_g) - E(t_{2g}) = 10D_q$, i.e., on whether Hund's rule for the maximal spin is satisfied in the crystal or not. If it is satisfied, the fourth electron fills the e_g^a-orbital and a so-called high-spin state is formed, since a low-spin state is found relatively seldom in compounds of 3d-metals. For ions with a $3d^5$ configuration (for example, the complex $[MnF_6]^{4-}$) in a high-spin state, three electrons are placed in the t_{2g}^a-orbitals and two in the e_g^a-orbitals. If the number of 3d-electrons is larger than five, the subsequent electrons fill the levels in the same order, starting with the t_{2g}^a-orbital, but with spins pointing in the opposite direction.

For $[NiF_6]^{4-}$, eight electrons are placed as shown in Figure 17. Two unpaired electrons in this complex are in the e_g^a-orbitals, which is consistent with the results of crystal field theory.

Now that the form of the wave functions and the corresponding energy level diagram have been found, the spin density at the fluorine nuclei can be calculated. For this purpose, we have to calculate the hyperfine interaction of the fluorine nucleus 3 with the electrons in the molecular orbitals. Since we are only considering terms for which the spin of the system is maximal ($S = 1$), the many-electron wave function can be written in the form of a Slater determinant made up of molecular orbitals. In this case, the full hyperfine field at the fluorine nucleus can be represented simply as a sum of fields contributed by each electron separately, i.e., as a sum of the ex-pectation values $(\Psi | H_e | \Psi)$ of the hyperfine field operator (1.1), taken for each filled molecular orbital separately. Using, for example, the explicit form of the mole-cular orbital $\Psi_{e(3Z^2 - r^2)}^a$, we find from (8.10) that the hyperfine field set up by an electron in this orbital has the form

$$H_{HFI}^z = -\frac{1}{2\gamma_n\hbar N_e}\left[\frac{\lambda_\sigma^2 a_{2p}(3\cos^2\theta_{\sigma Z} - 1)}{3} + \frac{\lambda_s^2 a_{2s}}{3}\right] +$$

$$+ N_e^{-1}(\phi_{e(3Z^2-r^2)}|H_e|\phi_{e(3Z^2-r^2)}). \tag{8.18}$$

The last term in (8.18) is usually small compared to the others, since the functions ϕ_e decrease fast with the distance and the contribution to the contact interaction (proportional to $|\phi_e(R)|^2$) is usually negligibly small (here R is the distance Ni—F). The dipole contribution, proportional to $(1/R^3)_{mean}$, is usually taken into account via a dipole field, which is easily found by assuming the paramagnetic ions around the fluorine as point dipoles. Therefore, dropping the last term, we can compute the spin density due to an electron in the $\Psi_{e(3Z^2-r^2)}^a$ orbital by equating (8.18) and (8.6). This gives

$$f_s^a = \frac{\lambda_s^2}{3N_e}, \quad f_\sigma^a = \frac{\lambda_\sigma^2}{3N_e}, \quad f_\pi^a = 0. \tag{8.19}$$

In order to find the full spin density at the nucleus, it is necessary to sum over all the orbitals.

It is obvious that of all the MO, only $\Psi_{e(3Z^2-r^2)}^a$ and $\Psi_{e(3Z^2-r^2)}^b$ will contribute to the HFI with nucleus 3, since the orbitals Ψ_t^a and Ψ_t^b contain no unpaired spins. At a first glance, it appears that the bonding orbital $\Psi_{e(3Z^2-r^2)}^b$ does not contribute to the HFI either, since it has two electrons with opposite spins. However, this is not so, as was shown by Watson and Freeman (1964) and Simanek and Sroubek (1964). The covalency parameters for orbitals of the same symmetry depend on the direction of the spins, if the system contains electrons with unpaired spins. Therefore we should assign to the parameters γ, and to all the normalization constants N_e in (8.14), an index indicating the direction of the electron spins. Then, for an electron of the $\Psi_{e(3Z^2-r^2)}^a$ orbital (spin "up"), we have, instead of (8.19), taking (8.17) into account,

$$f_s^a = \frac{(\gamma_{s\uparrow} + S_s)^2}{3N_{e\uparrow}} \quad \text{and} \quad f_\sigma^a = \frac{(\gamma_{\sigma\uparrow} + S_\sigma)^2}{3N_{e\uparrow}}. \tag{8.20}$$

Similarly, it is not difficult to find for the bonding orbitals $\Psi_{e(3Z^2-r^2)}^b$ for electrons with spin ↑

$$f_s^b = \frac{1}{3N_{e\uparrow}'}, \quad f_\sigma^b = \frac{1}{3N_{e\uparrow}''}, \tag{8.21}$$

and for electrons with spin ↓

$$f_s^b = -\frac{1}{3N_{e\downarrow}'}, \quad f_\sigma^b = -\frac{1}{3N_{e\downarrow}''}. \tag{8.22}$$

Using the explicit form of the normalization coefficients (8.14) and adding up the spin densities of the three electrons, we find (up to terms not higher than second

order in γ and S)

$$f_s = \frac{(\gamma_{s\downarrow} + S_s)^2}{3} \quad \text{and} \quad f_\sigma = \frac{(\gamma_{\sigma\downarrow} + S_\sigma)^2}{3}. \tag{8.23}$$

Thus, it turns out that the covalency (but not the overlap) of the $\uparrow$ electrons in anti-bonding orbitals is fully compensated by the covalency of the $\uparrow$ electrons in the bonding orbitals. This exact compensation is not accidental but follows from the properties of the many-electron functions, represented in the form of a determinant which is made up of one-electron orbitals.

Some remarks ought to be made on the relation between spin density and the parameter $N_e^{-1}\lambda^2/3$ in molecular orbitals. In general, this parameter represents the probability of finding the respective electron in a fluorine orbital. In the particular nickel complex being considered, the spin density turned out to be equal to this parameter because of the absence of orbital degeneracy in the system and the specific way in which the molecular orbitals are filled. Indeed, the parameter $N_e^{-1}\lambda^2/3$ (in our approximation) is not determined by the filling of the MO with electrons and it does not vanish even when there are no electrons with unpaired spins in the particular MO system. The spin density in this case is clearly zero and cannot be observed experimentally.

The spin density is not equal to this parameter when the maximal projection of the electron spin in the ground state is a nonconserved quantity. Therefore, in the general case, the spin density should be distinguished from the sum of the squares of the mixing coefficients of the atomic wave functions in the MO.

We have previously computed the HFI field for one of the simplest cases in which the orbital degeneracy of the electron system was lifted by an electric field in the crystal, i.e., the case of a locked orbital momentum. The calculations become more complicated when orbital momentum is not fully locked.

As a suitable example, let us briefly consider an octahedral complex (see Figure 16) containing a cobalt atom, i.e., $M \equiv Co$.

Since we are dealing with a weakly covalent crystal and neglect all correlation between bonds formed by different electrons, the results of crystal field theory can be used for the solution of the problem. In this approximation, it is assumed that the complex consists of a divalent Co^{2+} ion surrounded by six fluorine ions. The many-electron wave functions for Co^{2+} in the octahedron should be found and expressed in terms of the one-electron $3d$-functions. In order to allow for the covalency, the "pure" one-electron functions of the metal are then replaced by the appropriate molecular orbitals. Detailed calculations for such a complex can be found in the work of Tornley et al. (1965) and for a more complicated case in the article of Petrov and Nedlin (1967). Here we shall only consider some features which arise when the spin density at the fluorine nuclei is being calculated.

In accordance with the Hund rules, the free Co^{2+} ion is in a ^4F-state (spin $S = \frac{3}{2}$).

A sufficiently weak crystal field splits this fivefold degenerate orbital state into three levels, the lowest of which is an orbital triplet. (If spin is taken into account, this level is twelvefold degenerate). Spin-orbit interaction partially lifts the twelvefold degeneracy and the ground state becomes a Kramers doublet which can be described by means of some effective total momentum $J = \frac{1}{2}(J^z = l^z + S^z$, where l is the effective orbital momentum). Since only J^z has eigenvalues in this state, and not S^z (S^z does not commute with the spin-orbit interaction operator), levels with $J^z = \frac{1}{2}$ and $J^z = -\frac{1}{2}$ are superpositions of states with different projections S^z (for $S = \frac{3}{2}$).

The wave functions Ψ_{J, J^z} describing these levels with $J = \frac{1}{2}$ and $J^z = \pm \frac{1}{2}$ are thus no longer represented by pure determinants, and the spin density at the fluorine nuclei cannot be obtained by simply adding up the squares of the mixing coefficients of the fluorine atomic wave functions. In this case, we use the many-electron functions Ψ_{J, J^z} to compute the total hyperfine field at the nucleus as the expectation value of the operator H_e^{tot}, which is a sum of one-electron operators of the type (1.1), i.e., $H_e^{\text{tot}} = \sum_i H_e^i$. If, for the sake of clarity, we limit ourselves only to the contact interaction with electrons in antibonding orbitals, and ignore the interaction between the ^4F- and ^4P-terms, we obtain

$$H_{\text{HFI}}^z = (\Psi_{1/2, 1/2} \, | \, H_e^{\text{tot}} \, | \, \Psi_{1/2, 1/2}) = -\frac{A_s}{\gamma_n \hbar}(\Psi_{1/2, 1/2} \, | \, S^z \, | \, \Psi_{1/2, 1/2}), \qquad (8.24)$$

where

$$A_s = \frac{a_{2s} N_e^{-1} \lambda_s^2}{9}. \qquad (8.25)$$

Further,

$$(\Psi_{1/2, 1/2} \, | \, S^z \, | \, \Psi_{1/2, 1/2}) = \bar{S}^z = \frac{5}{6}. \qquad (8.26)$$

As a result, we obtain for the spin density

$$f_s = \frac{2A_s \bar{S}^z}{a_{2s}} = \frac{5}{9} \frac{\lambda_s^2}{3N_e}. \qquad (8.27)$$

We see from (8.27) that in this case, the spin density is not equal to $\lambda_s^2/3N_e$, i.e., in contrast to the expression obtained for $[NiF_6]^{4-}$, the polarization of the fluorine shell is not determined by the squares of the mixing coefficients.

Another interesting feature of the hyperfine field in this case is its anisotropy, which is evident for the contact interaction, if the admixture of excited configurations due to the spin-orbit interaction is taken into account (Hall et al., 1963). Therefore, in order to generalize (8.2) to this case, two different values ($\parallel$ and $\perp$) have to be assigned to the parameter A_s, since

$$A_{s\perp} \neq A_{s\parallel}.$$

In a number of experiments, which we shall discuss in Sec. 4, spin density was studied at the nuclei of nonmagnetic metal ions in paramagnetic fluoride crystals with a perovskite-type structure. The general formula is AMF_3, where $A \equiv Tl$, Rb, K, etc., $M \equiv Mn$, Ni, Co, etc. (the unit cell of this structure is shown in Figure 18). We shall now consider the origin of spin density at the nuclei A.

As the symmetry of the surroundings of the nuclei A is cubic, the dipole HFI vanishes and the entire observed HFI comprises contact interaction due to the s-shell polarization of the ion A. We shall not specify here exactly which s-shell makes the predominant contribution.

The calculation can again be carried out by the method of molecular orbitals for the augmented complex $[A_8MF_6]^{4+}$, including the eight A ions. The sought molecular orbitals, in addition to the e_g- and t_{2g}-orbitals of the paramagnetic ion and the corresponding combinations (8.12) and (8.13) of the fluorine orbitals, should also include combinations of the A ion s-orbitals which transform as the e_g- or t_{2g}-functions. Using standard group-theoretical methods, we find that the combinations of the s-functions of A ions (atoms) may belong to the irreducible representations a_{1g}, a_{2u}, t_{1u}, t_{2g}. The A atom s-functions thus can mix only with the t_{2g}-orbitals but not with the e_g-orbitals of the central ion. It is readily seen that the combinations of the A atom s-functions from the t_{2g}-representation have the form

$$\eta_{s(XY)} = \frac{1}{\sqrt{8}}(\Sigma_1 + \Sigma_2 + \Sigma_5 + \Sigma_6 - \Sigma_3 - \Sigma_4 - \Sigma_7 - \Sigma_8),$$

$$\eta_{s(YZ)} = \frac{1}{\sqrt{8}}(\Sigma_1 + \Sigma_3 + \Sigma_5 + \Sigma_7 - \Sigma_2 - \Sigma_4 - \Sigma_6 - \Sigma_8), \qquad (8.28)$$

$$\eta_{s(XZ)} = \frac{1}{\sqrt{8}}(\Sigma_1 + \Sigma_4 + \Sigma_5 + \Sigma_8 - \Sigma_2 - \Sigma_3 - \Sigma_6 - \Sigma_7),$$

where Σ_i is the s-function of the i-th A atom.

Instead of the former (8.11), we can now find new molecular orbitals which also include these combinations:

$$\Psi_t^a = N_t^{-1/2}(\phi_t - \lambda_\pi \chi_\pi + \lambda \eta_s),$$
$$\Psi_t^b = N_t'^{-1/2}(\chi_\pi + \gamma_\pi \phi_t + \beta \eta_s), \qquad (8.29)$$
$$\Psi_t^A = N_t''^{-1/2}(\eta_s + \nu \phi_t + \mu \chi_\pi).$$

Each of these three orbitals stands for a triplet of molecular orbitals transforming as XY, YZ, XZ. In all, there are nine t_{2g}-orbitals.

Here N_t, N_t', N_t'' are new normalization coefficients whose explicit form is not given here. In order to ensure orthogonality, the following conditions must be fulfilled:

$$\gamma_\pi + S_{MF} - \lambda_\pi \simeq 0,$$

$$\beta + S_{AF} + \mu \simeq 0, \tag{8.30}$$

$$\nu + S_{MA} + \alpha \simeq 0,$$

where S_{MF}, S_{AF}, S_{MA} are group overlap integrals of the orbitals (M is the paramagnetic ion). The relative energy location of the orbitals Ψ_t^a and Ψ_t^b remains the same as before, whereas the orbital Ψ_t^A may lie above or below, depending on the particular A ion s-orbitals it includes.

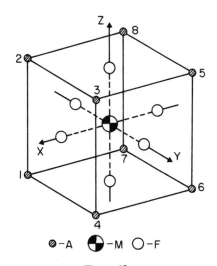

Figure 18

Part of the unit cell of a perovskite-type structure.

Let us consider a concrete example. The A atom is rubidium with its outer (valency) $5s$-shell; the paramagnetic ion is manganese. In the complex $[\mathrm{Rb_8MnF_6}]^{4+}$ the orbital Ψ_t^b is completely filled, the orbital Ψ_t^a is half-filled (spins $\uparrow$), and the orbital Ψ_t^A has no electrons. Here, as for the nickel complex, the many-electron state is described by a single determinant, and the hyperfine interaction of the rubidium nucleus is therefore simply the sum of all the HFI with the electrons in the molecular orbitals, while the spin density at the rubidium nucleus is the sum of the squares of the coefficients of η_s in electron-containing orbitals. If we confine ourselves to electrons in the Ψ_t^a orbital only, then

$$A_s^{\mathrm{Rb}} = \frac{3}{5} \frac{N_t^{-1}\alpha^2}{8} a_{5s} \tag{8.31}$$

and

$$f_s^{\mathrm{Rb}} = \tfrac{3}{8} N_t^{-1}\alpha^2 . \tag{8.32}$$

If we also take into consideration electrons in the Ψ_t^b orbitals, $N_t^{-1}\alpha^2$ in (8.31) and (8.32) must be replaced by $N_{t\uparrow}^{-1}\alpha_\uparrow^2 + N_{t\uparrow}'^{-1}\beta_\uparrow^2 - N_{t\downarrow}'^{-1}\beta_\downarrow^2$. It is interesting to note that if

$$N_{t\downarrow}'^{-1}\beta_\downarrow^2 > N_{t\uparrow}^{-1}\alpha_\uparrow^2 + N_{t\uparrow}'^{-1}\beta_\uparrow^2, \qquad (8.33)$$

the spin density and the constant A_s^{Rb} will both be negative (according to data of Kellog and Millman (1946), $a_{5s} = 0.11399\ \mathrm{cm}^{-1}$). Similar calculations for the complex $[Rb_8CoF_6]^{4-}$ were published by Petrov and Nedlin (1967).

3. Discussion of Experimental Work on the NMR of Fluorine

This section presents the results of a number of experimental studies devoted to spin density. Spin density measurements are usually carried out in the paramagnetic phase. This has certain methodological advantages, since measurements in the paramagnetic phase can be carried out with ordinary standard spectrometers. Moreover, in the paramagnetic phase, the HFI constants and the spin density can be determined quite unambiguously, since additional difficulties associated with the existence of a complicated and fairly uncertain magnetic structure in the ordered phase are usually excluded. There are, however, exceptions: for example, during measurements in $RbNiF_3$, about which more will be said below, a complex magnetic structure was discovered also in the paramagnetic phase.

An observable quantity in NMR experiments (besides line shape and relaxation time) is the magnetic field at which a signal is recorded. The resonance frequency is $\nu_0 = (\gamma_n/2\pi)\,H_{ef}$, where H_{ef} is some total magnetic field including in the general case the external (H), the dipole (H_{dip}), and the hyperfine (H_{HFI}) fields (see Chapter II, Sec. 1). In the simplest case, H_{dip} and H_{HFI} in the paramagnetic phase are usually small compared with H. Accordingly, in the first approximation, only the longitudinal components of H_{dip} and H_{HFI} contribute to H_{ef}. Indeed, if the direction of H is chosen along z,

$$H_{ef} =$$

$$= \sqrt{(H + H_{dip}^z + H_{HFI}^z)^2 + (H_{dip}^x + H_{HFI}^x)^2 + (H_{dip}^y + H_{HFI}^y)^2} \approx H + H_{dip} + H_{HFI},$$

$$(8.34)$$

where we have omitted the superscript z of the longitudinal components of H_{HFI} and H_{dip}. However, even in this case, because of the tensor character of the HFI and the g-factor, not only S^z but also the S^x and S^y components of the paramagnetic-ion spin contribute to the longitudinal component H_{HFI}^z.

For finite temperatures T, the quantum-mechanical mean $\bar{S}^z$ is replaced by the thermodynamic mean $\langle S^z \rangle$ in expression (8.2) which defines the hyperfine field (in the

axisymmetric case). In other words,

$$H_{\text{HFI}} = - (\gamma_n \hbar)^{-1} \sum_j [A_{sj} + A_{pj} (3 \cos^2 \theta_{\sigma z} - 1)] \langle S_j^z \rangle, \tag{8.35}$$

where

$$\langle S_j^z \rangle = \frac{\text{Tr} [S_j^z \exp (- \beta \mathcal{H}_s)]}{\text{Tr} [\exp (- \beta \mathcal{H}_s)]}. \tag{8.36}$$

Here $\mathcal{H}_s$ is the Hamiltonian which, in general, incorporates the interaction of the ion with the external magnetic and the crystal fields, and also the exchange interaction, $\beta = 1/\kappa T$. The summation in (8.35) is carried out over the nearest-neighbor paramagnetic ions.

For ions in the S-state in the paramagnetic phase

$$\langle S_j^s \rangle = \frac{\chi_m H}{N_A \gamma_e \hbar}, \tag{8.37}$$

where χ_m is the molar susceptibility, while N_A is the Avogadro number. (We recall that in our notation, $\gamma_e < 0$.) In the general case, when an orbital momentum admixture contributes to the susceptibility, $\langle S_j^z \rangle$ has a more complicated form. For Ni^{2+} in octahedral surroundings (Shulman, 1961 a, b)

$$\langle S_j^z \rangle = \frac{H \left(\chi_m - N_A \mu_B^2 \frac{\Delta g}{\lambda} \right)}{N_A \gamma_e \hbar}, \tag{8.38}$$

where $\Delta g = 2.0023 - g$, and λ is the spin-orbit interaction constant. An expression for the mean spin of the Co^{2+} ion can be found, e.g., in Tsang (1964) and Kamimura (1966).

Relation (2.4), with $\langle \mu_i \rangle = \chi_m H/N_A$, is valid for the dipole field H_{dip}. The NMR frequency ω_n for an axisymmetric surrounding of the nucleus takes the form

$$\omega_n = 2\pi v_0 = \gamma_n (H + H_{\text{dip}}) - \frac{1}{\hbar} \sum_j [A_{sj} + A_{pj} (3 \cos^2 \theta_{\sigma z} - 1)] \langle S_j^z \rangle. \tag{8.39}$$

The corresponding expression for ω_n for lower symmetry can be found in the work of Marshall and Stuart (1961), Shulman and Jaccarino (1957) and Shulman (1961).

In practice, the local field $\Delta H = H_{\text{HFI}} + H_{\text{dip}}$ is determined from the shift of the NMR line in the crystal with respect to some diamagnetic standard sample containing similar resonating nuclei. If neither A_s nor A_p vanish, the H_{HFI} shift has both an isotropic and an anisotropic component.

In polycrystalline samples, the anisotropic components of the local field undergo angular averaging and produce a single line of a specific irregular shape. A typical curve is shown in Figure 19 for $NaNiF_3$ (Petrov, 1965). The anisotropic components

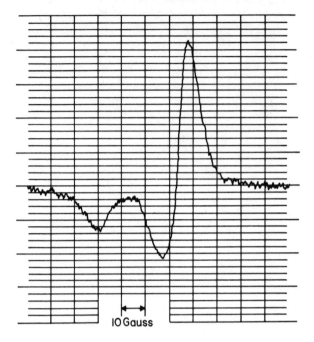

Figure 19

The NMR line shape of F^{19} in polycrystalline $NaNiF_3$ at room temperature ($f_0 = 14.7$ MHz).

of the local field in this case can be computed by a method worked out in electron paramagnetic resonance studies for the calculation of the anisotropic component of the g-factor in polycrystals from the EPR line shape (Blyumenfel'd et al., 1962; Chirkov and Kokin, 1960).

It is seen from (8.39) that, by measuring the local fields at the nuclei in the paramagnetic phase, we can in principle study the temperature variation of the sublattice magnetizations (more precisely of $\langle S^z \rangle$) above the magnetic ordering temperature, determine the HFI constants, and thence the spin density and the covalency, and also (from the character of the angular dependence) the bond angles between the paramagnetic and diamagnetic ions.

The measurement of the fluorine NMR in $RbNiF_3$ (Smolenskii et al., 1967) is an example of the study of the sublattice magnetizations in the paramagnetic phase. In $RbNiF_3$, there are two nonequivalent positions of the Ni^{2+} ion, forming two magnetic sublattices a and f (Figure 20). There are also two nonequivalent positions of the fluorine ions (h and k).

In Figure 21a, dots give the values of the inner field at the fluorine nuclei $H \parallel C$. For the k nuclei, virtually the entire ΔH is due to the isotropic hyperfine field, while for the h nuclei there is a small contribution from the anisotropic component. As

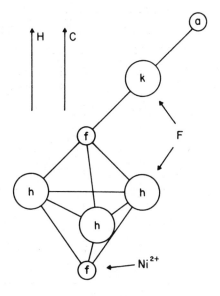

Figure 20

A part of the unit cell of RbNiF$_3$.

we see from the figure, the local field ΔH at the fluorine nuclei in the h position predictably increases with the lowering of the temperature, since the magnetization increases as the Curie point is approached. However, the local field at the k nuclei conversely decreases and even reverses its sign as the temperature is lowered. Such an "anomaly" can be explained in the following way.

From crystallographic data (Figure 20) it follows that the h ions of fluorine are bonded only to the nickel ions of the f sublattice (the number of which is double the number of ions in the a sublattice). Therefore, the isotropic hyperfine field for the h nuclei is

$$\Delta H_{is}^h = -\frac{2}{\gamma_n \hbar} A_{sh}^f \langle S_f^z \rangle \tag{8.40}$$

and the temperature variation of ΔH_{is}^h is fully determined by $\langle S_f^z \rangle$. At the same time, the k ions of fluorine are bonded to nickel ions of both sublattices, and therefore for the k nuclei

$$\Delta H_{is}^k = -\frac{1}{\gamma_n \hbar} \left[A_{sk}^f \langle S_f^z \rangle + A_{sk}^a \langle S_a^z \rangle \right]. \tag{8.41}$$

Since all the HFI constants for fluorine must be positive, the sign of ΔH_{is}^k is reversed only if the sign of $\langle S_a^z \rangle$ changes and

$$\left| A_{sk}^a \langle S_a^z \rangle \right| > \left| A_{sk}^f \langle S_f^z \rangle \right|. \tag{8.42}$$

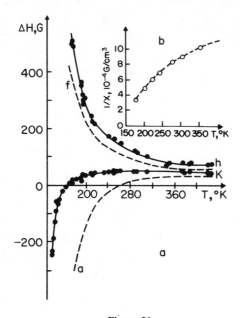

Figure 21

The temperature dependence of ΔH for $H \parallel C$.

a   The dashed lines show the temperature dependence of $\langle S_f^z \rangle$ and $\langle S_a^z \rangle$ in arbitrary units.   *b*   The calculated temperature dependence of $1/\chi$ is shown by the dashed line. The dots are the experimental values taken from Smolenskii et al. (1966) and corrected for the Van Vleck paramagnetism.

The temperature dependence of $\langle S_f^z \rangle$ and $\langle S_a^z \rangle$ shown (in arbitrary units) by the dashed lines in Figure 21a explains the behavior of the observed local fields at the fluorine nuclei. Already at temperatures some 130° above T_C the spin of one sublattice is directed against the spin of the other sublattice, i.e., the specimen exhibits a distinctive ferrimagnetism in a magnetic field above T_C (139°K).

Choosing appropriate values for the HFI constants in (8.40) and (8.41), we can find the absolute values of $\langle S_a^z \rangle$ and $\langle S_f^z \rangle$ and then calculate the temperature dependence of susceptibility. The results of such calculations are shown by the dashed line in Figure 21b. The dots are the experimental values of the reciprocal susceptibility (corrected for the temperature-independent paramagnetism). Thus, comparison with susceptibility supports the conclusion regarding the existence of "ferrimagnetism" above T_C.

Highly comprehensive and interesting experimental investigations of spin density have lately been carried out in paramagnetic fluorine compounds with a perovskite-type structure (see Figure 18). It is very convenient to take measurements on fluorine nuclei because their gyromagnetic ratio is large and the natural isotope content of ^{19}F is 100 per cent. The interpretation of the results is quite unambiguous, since the

Compound (reference)	Lattice constants, Å	Fluorine				Nuclei of ion A	
		A_s (10^4 cm^{-1})	f_s, %	A_p, 10^4 cm^{-1}	f_p, %	A_s (10^4 cm^{-1})	f_s, %
KMnF$_3$ (a)	$a = 4.19$	16.26	0.53	0.18			
TlMnF$_3$ (b)	$a = 4.25$	15.3	0.51			−3.5	−0.03
RbMnF$_3$ (c, j, k)	$a = 4.25$	15.7	0.52	0.4 (j)		−0.12	−0.052
NaMnF$_3$ (d)	$a = c = 3.997$ $b = 3.992$	15.8	0.52				
KNiF$_3$ (e)	$a = 4.04$	33.9	$\begin{cases} 0.451 \\ (0.538) \end{cases}$	8.3	3.78		
NaNiF$_3$ (f)	$\begin{cases} a = 5.361 \\ b = 5.524 \\ c = 7.688 \end{cases}$	41	0.55	9	4.2	∼ 0	∼ 0
KCoF$_3$ (g), (h)	$a = 4.069$	25	$\frac{\lambda_s^2}{3N_e} = 0.51$	9			
RbCoF$_3$ (i)	$a = 4.116$	19	$\frac{\lambda_s^2}{3N_e} = 0.4$	6		−0.093	−0.013

Note:

1. The values of f_s (F) are not corrected for the $1s$–$2s$ interaction mentioned in the text (see below); for KNiF$_3$ the corrected value is given in parentheses.

2. The data for TlMnF$_3$ borrowed from Petrov and Smolenskii (1965) are recalculated in accordance with our notation.

3. The values of $\lambda_s^2/3N_e$ in the cobalt compounds are presented without the corrections introduced in Tsang (1964).

4. *References:* (a) Shulman and Knox (1960b), (b) Petrov and Smolenskii (1965), (c) Petrov et al. (1965), (d) Petrov and Moskalev (1968), (e) Shulman and Sugano (1963), (f) Petrov and Smolenskii (1966), (g) Tsang (1964), (h) Shulman and Knox (1965), (i) Petrov and Nedlin (1967), (j) Payne et al. (1965), (k) Wolker and Stevenson (1966).

fluorine ion has only one p-shell ($2p$) with electrons, in distinction to Cl, Br, Se, S, and other ligands where the field at the nucleus depends on the spin density of different p-shells whose contributions cannot be separated experimentally. Moreover, the fluorine compounds mentioned above contain other ions for which NMR can be conveniently observed.

The simple and highly symmetrical structure of these crystals enables us to make an effective use of the molecular orbitals method.

The table on page 150 lists experimental data on local fields and corresponding spin density estimates for a number of fluorides with a perovskite structure.

Using the spin density data, it is possible to calculate the degree of covalency. Shulman and Sugano (1963) obtained for the Ni—F bond in $KNiF_3$ $\gamma_\sigma \simeq 0.3$, while Freeman and Watson (1961) obtained for the Mn—F bond in $KMnF_3$ $\gamma_s \simeq 0.03$. Close estimates can be obtained for $NaNiF_3$ and the manganese compounds listed in the table. Note that from NMR measurements we can calculate the covalency of the different electron groups ($2s$ or $2p$) entering the chemical bond.

Figure 22 shows the angular dependence of the resonance magnetic field H_{ef} for fluorine nuclei in single crystals of $KMnF_3$, $KNiF_3$, and K_2NaCrF_6 (Shulman and Sugano, 1963). The resonance field H_{ef} for the unshifted fluorine NMR line, i.e., for the NMR line in a diamagnet, is 14.979 kOe ($\nu_0 = 60$ MHz). The figure shows that the NMR line in a paramagnet is strongly displaced and the shift indeed has an isotropic and an anistropic component. The calculated values of H_{dip} for the crystals are shown by the dashed lines. In $KMnF_3$ the anisotropic shift is almost fully accounted for by the dipole field, i.e., there is very little anisotropic hyperfine interaction. In $KNiF_3$ and K_2NaCrF_6, both the dipole and the hyperfine fields are responsible for the anisotropy of the shifts.

We see that $A_p = A_\sigma - A_\pi$ is considerably smaller for the manganese compounds than for the nickel compounds. Shulman and Sugano (1963) inferred from this that the σ- and π-bonds in manganese are approximately of the same intensity, although, generally speaking, this difference may be small also when A_σ and A_π are both small quantities.

Shulman and Sugano (1963) also made an interesting comparison between the results for $KNiF_3$ and K_2NaCrF_6. In $KNiF_3$, the unpaired electrons are in the Ψ_e^a molecular orbital and form σ-bonds only, so that $A_\sigma \neq 0$ and $A_\pi = 0$. In K_2NaCrF_6 the unpaired electrons occupy the Ψ_t^a molecular orbitals and form π-bonds only. Therefore in the latter case we should have $A_\sigma = 0$ and $A_\pi \neq 0$. The observed angular dependences of the hyperfine fields for $KNiF_3$ and K_2NaCrF_6 are very much alike, but opposite in phase (Figure 22). Thus, experiments confirm that $A_p \simeq A_\sigma$ in $KNiF_3$, and $A_p \simeq -A_\pi$ in K_2NaCrF_6. Moreover, almost no isotropic hyperfine field is observed in K_2NaCrF_6. This fits the fact that the s-electrons of fluorine do not form bonds with the t_{2g}-electrons of chromium.

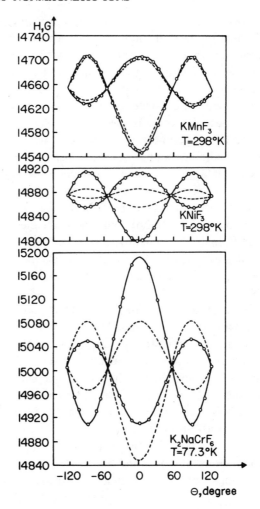

Figure 22

The angular dependence of the local field at the F^{19} nuclei in $KNiF_3$, $KMnF_3$, and K_2NaCrF_6.

The dashed lines represent the calculated dependence for the dipole field. θ is the angle between H and the [001] axis.

4. NMR of Metallic Nonmagnetic Ions

After the investigation of spin density at ligand nuclei, the next step was spin density measurements at the nuclei of nonmagnetic cations. Some of the early works in this direction dealt with the NMR frequency shift in the paramagnetic compound $LiMnPO_4$ (NMR observations at the nuclei of Li and P (Mays, 1963) and field

measurements at the Ga nuclei in the ferrite–garnet $Y_3Fe_{5-x}Ga_xO_3$ (Streever and Uriano, 1965). However, the interpretation of the observed effects in these compounds is difficult because of the complexity of their crystal structure. The AMF_3 crystals with perovskite structure are easier to handle in this respect. In previous studies (see references in the table), the fields were measured not only at the fluorine nuclei but also for A = Tl, Rb, Na in $TlMnF_3$, $RbMnF_3$, $NaNiF_3$, and $RbCoF_3$. In crystals with M = Mn or Co fields were observed at the nuclei of the A ions, whereas no field was detected at the Na nuclei in $NaNiF_3$ (see table). The observed local fields had only an isotropic component and were directed opposite to the external field. Because of the cubic symmetry of the surroundings, the dipole fields at the A nuclei of the paramagnetic ions inside the Lorentz sphere and the anisotropic hyperfine fields all vanish. The sum of the demagnetizing field and the Lorentz field is considerably less than the observed field. Therefore, the observed isotropic field at the A nuclei is due to the presence of spin density at these nuclei, i.e., to the polarization of the s-shells in the A ions. The spin density is calculated from experimental data exactly as in the previous section, but in this case the Hamiltonian contains no angular terms. The experimental spin density values are presented in the table.

In calculations of spin density at nuclei, the entire spin density is attributed to the polarization of one outer s-shell. For example, in the Tl^+ ions, the $6s$-shell is assumed to be polarized (the HFI constant for this shell is $a_{6s} = 5.9$ cm^{-1} (Fermi and Segre, 1933)), whereas for rubidium nuclei the spin density is assumed to be associated with the partial presence of an unpaired electron in the outer (valence) $5s$-shell of the rubidium ion. The contribution from deeper lying s-shells can undoubtedly change the tabulated absolute values of the spin density; however, this is not very important when we are interested in the relative values of the spin density for the same ion A in various crystals.

The different signs of f_s for fluorine and the nuclei A indicate that the spin density goes through zero somewhere in the space between the fluorine nucleus and A, i.e., there are characteristic spatial oscillations of spin density.

Note that f_s^{Rb} decreases on passing from $RbMnF_3$ to $RbCoF_3$. It follows from the calculations in Sec. 3 that only the unpaired t_{2g} electrons contribute to spin density in rubidium. In $RbMnF_3$, there are three such electrons while in $RbCoF_3$ there is approximately one; therefore the magnitude of f_s^{Rb} in $RbCoF_3$ is less.

If only the electrons in antibonding orbitals are taken into account and it is assumed that $N_t^{-1}\alpha^2$ from (8.29) is the same for $RbCoF_3$ and $RbMnF_3$, it can be shown (Petrov and Nedlin, 1967) that

$$\frac{(f_s^{Rb})_{RbCoF_3}}{(f_s^{Rb})_{RbMnF_3}} \approx 0.202. \tag{8.43}$$

Experimental data indicate that this ratio is equal to 0.26, i.e., there is a fairly good agreement between theory and experiment. Experiments confirm convincingly

that the spin density at the rubidium nuclei is due to the t_{2g}-electrons. This conclusion is also supported by the fact that there is no spin density at Na in NaNiF$_3$, since, in this case, there are no unpaired t_{2g}-electrons.

5. Remarks on the Calculation of Spin Density

It is necessary to mention a number of effects and corrections which, when taken into account, can markedly change the spin density and the covalency calculated from experiment.

1. The influence of deeper lying shells. We have so far assumed that the major contribution to spin density at the nuclei came from one shell, namely the outer s-shell. For example, for fluorine only the 2s-shell was considered. Watson and Freeman (1961), and also Marshall and Stuart (1961) considered the contribution from the 1s-shell. It was shown that the resulting correction to f_s, which is proportional to the product $|\psi_{2s}(0)\psi_{1s}(0) \cdot S_{1s} \cdot S_{2s}|$, can reach 15–20% (here S_{1s} and S_{2s} are the overlap integrals (8.15) for the 1s- and 2s-shells, respectively).

2. Interaction between 2s- and 2p-electrons. Terms of the form $\gamma_{\sigma s}\chi_s$ and $\gamma_{s\sigma}\chi_\sigma$ in general should be taken into account in (8.10). However, their contribution is small and their inclusion does not affect the spin density in the first approximation (Watson and Freeman, 1964). Moreover, the exchange polarization of the 2s-shell interacting with the polarized 2p-shells leads to an additional isotropic HFI, which is caused by the polarization of the 2s-orbitals.

It would be interesting to take this circumstance into account in the analysis of spin density at the A nuclei. However, numerical estimates are difficult to obtain.

3. Admixture of excited states. The admixture of excited states as a result of spin-orbit or configurational interactions can introduce corrections to A_s and A_p reaching 10% for Co-containing crystals (Tornley et al., 1965). Corrections to the HFI constants due to the spin-orbit interaction are also important for nickel compounds (Shulman and Sugano, 1963).

Some inaccuracies which arise in calculations of spin density from experiments should also be mentioned. One of them is the calculation of the dipole fields. Errors in the calculations of H_{dip} can arise if the magnetic ions are assumed to be point dipoles and their spatial extent is ignored (Marshall and Stuart, 1961; Shulman and Sugano, 1963). Another inaccuracy lies in the computation of $\langle S^z \rangle$. Thus calculations of $\langle S^z \rangle$ in cobalt compounds require values of the g-factor, spin-orbit interaction constants, the exchange integral and the Racah parameter, not all of which are known with sufficient accuracy. Unfortunately, this uncertainty in the parameters can lead to an uncertainty in the HFI constants exceeding the experimental errors.

In addition to the above effects, there are also other reasons associated with a fundamentally different approach to the interpretation of the observed hyperfine

fields. As previously stated at the beginning of the chapter, the hyperfine field at the nuclei of nonmagnetic ions can be accounted for by considering excited states in which the ion system M^{2+}—F^- makes a transition to the state M^+—F. In this interpretation, the field at the fluorine nucleus is due to the presence of an unpaired electron in the F atom (Sugano and Tanabe, 1965), and the spin density in the 2s-shell is proportional to A_s/a_{2s}^a, where a_{2s}^a is the HFI constant for the fluorine atom, and not for the ion, as before. The difference between the HFI constants for the fluorine ion and atom reaches 20–30% and constitutes still another source of uncertainty in the calculation of spin densities.

In calculations according to the configurational interaction method (Rimmer, 1965; Hubbard et al., 1966) effects associated with electron transitions from the fluorine ion shell to an empty 4s-shell of a paramagnetic ion are also considered. According to Hund's rule, an electron with a spin parallel to the spin of the electrons in the 3d-shell has a high transition probability. These transitions constitute still another source of polarization of the 2s- and 2p-shells of fluorine, and introduce an additional uncertainty in the determination of such parameters as the covalency.

It is difficult to take all these effects into account in practical work, and the values presented in the literature generally do not include these corrections. The absolute values of spin density and covalency obtainable in this way are therefore not particularly accurate.

CHAPTER IX

Nuclear Magnetoacoustic Resonance (NMAR)

1. Resonant Ultrasound Absorption Coefficient

As has been shown in Chapter V (Sec. 7) and Chapter VI (Sec. 3), nuclear spins and lattice vibrations may interact via either one of three mechanisms. First, there is the indirect interaction between the nuclear spins and the lattice via spin waves. Because of magnetoelastic coupling, lattice vibrations excite vibrations of the electron spins, which, in their turn, act on the nuclear spins through hyperfine coupling. Second, there is the modulation of the hyperfine or the dipole interaction between the nuclear and the electron moments. And, last, the quadrupole mechanism, produced by the change (or appearance, in the case of nuclei with cubic surroundings) of the quadrupole interaction due to lattice deformations.

Direct (one-phonon) processes of spin-lattice relaxation caused by these mechanisms frequently prove ineffective, because the energy density of thermal phonons at NMR frequencies is very small. It is therefore of great interest for the investigation of the mechanisms mentioned to excite nuclear spins by ultrasound vibrations, which can concentrate considerable energy in a narrow frequency interval (Abragam, 1963). Moreover, in ultrasound experiments, the various interaction mechanisms can sometimes be separated, since for different polarizations, different directions of propagation and different ultrasound intensities (all of which can be controlled), different interaction mechanisms will in general manifest themselves.

The ultrasonic lattice vibrations generate a variable magnetic field of the same frequency at the nuclei, which is proportional to the components of the symmetrical strain tensor $u_{\alpha\beta}$ (or the skew-symmetrical rotation tensor $\varepsilon_{\alpha\beta}$). The explicit form of these fields for various mechanisms was presented in the discussion on spin-lattice relaxation (see Chapter V, Sec. 7 and Chapter VI, Sec. 3). When the ultrasound

frequency is close to the NMR frequency, these magnetic fields will induce resonance transitions between the quasi-Zeeman levels of the nuclear spins. Thus resonant absorption of ultrasound will be observed at the NMR frequency. Such a phenomenon is known as nuclear magnetoacoustic resonance (NMAR).

The specific features of NMAR in ferro- and antiferromagnets are closely associated with the specific nuclear spin-lattice interaction mechanisms in these materials. Since the magnetoelastic interaction is one of such specific mechanisms, it will be given the main consideration below. However, it will not be difficult to show how the results change when other mechanisms are applied.

Let us consider a monochromatic sound wave of the form

$$u = u_0 e \sin(qr - \Omega t), \tag{9.1}$$

where u_0, e, q, and Ω are the amplitude, the unit polarization vector, the wave vector, and the lattice vibration frequency, respectively. The transverse (with respect to the nuclear spin quantization axis) components of the variable magnetic field set up by these vibrations can always be written in the following form (see below):

$$\delta H^\pm = H_1^\pm \cos(qr - \Omega t), \tag{9.2}$$

where the explicit form of the field amplitude $H_1^\pm$ is determined by the actual type of interaction. The average energy absorbed per second by the nuclear spin system in this field can be calculated from the known expression

$$\mathscr{P} = \tfrac{1}{2}\Omega \chi''(q, \Omega) H_1^2 \tag{9.3}$$

($H_1^2 = H_1^+ H_1^-$), where $\chi''(q, \Omega)$ is the imaginary part of the complex magnetic susceptibility of the nuclear spin system in the direction of the field H_1 (if the power $\mathscr{P}$ is related to unit volume, the susceptibility must correspondingly be taken for unit volume).

If the spatial correlation in the motion of the nuclear spins is ignored,* it can be assumed that there is no spatial dispersion of magnetic susceptibility and thus

$$\chi''(q, \Omega) \simeq \chi''(\Omega) \tag{9.4}$$

is the corresponding susceptibility for NMR in a homogeneous radiofrequency field.

The coefficient of sound absorption is defined as the ratio of the dissipated energy to double the mean sound energy flux (Landau and Lifshits, 1965):

$$\alpha^\lambda = \frac{\mathscr{P}_\lambda}{2c_s^\lambda \bar{E}}, \tag{9.5}$$

---

* For ferro- and antiferromagnets this apparently can be done when the correlation due to the indirect (Suhl–Nakamura) interaction between the nuclear spins via spin waves is small, or, in other words, when the existence of nuclear spin waves still may be ignored (see Chapter IV).

where λ is a polarization index and

$$\bar{E} = \frac{1}{V} \rho \int \overline{\dot{u}^2} \, dV = \frac{1}{2} \rho u_0^2 \Omega^2 \, . \tag{9.6}$$

Introducing the NMR line shape function $f(\Omega)$ by the known relation

$$\chi''(\Omega) = \tfrac{1}{2} \pi \chi_n \Omega f(\Omega) \, , \tag{9.7}$$

where (at $\kappa T \gg \hbar \omega_n$)

$$\chi_n = \frac{I(I+1)(\gamma_n \hbar)^2 N_n}{3\kappa T} \tag{9.8}$$

is the static susceptibility of N_n nuclei, we obtain from (9.5), using (9.3) and (9.7),

$$\alpha^\lambda = \frac{\pi \chi_n f(\Omega) H_1^2}{4 c_s^\lambda \rho u_0^2} \, . \tag{9.9}$$

Thus, for the calculation of the absorption coefficient, it remains to substitute in the last expression the appropriate value of the amplitude H_1 of the effective magnetic field produced by the ultrasound wave.

2. NMAR in Ferromagnets

In ferromagnets, the field $\delta H^\pm$ set up by lattice vibrations, assuming the magneto-elastic mechanism, is determined by (5.87). For a monochromatic wave (9.1), $u_{\alpha\beta}(k)$ by (5.76) has the form

$$u_{\alpha\beta}(k) = \tfrac{1}{4} u_0 (e_\alpha q_\beta + e_\beta q_\alpha) [\delta(k+q) e^{-i\Omega t} + \delta(k-q) e^{i\Omega t}] \, , \tag{9.10}$$

and the analogous expression for $\varepsilon_{\alpha\beta}(k)$ is obtained from (9.10) by the substitution of $(e_\alpha q_\beta - e_\beta q_\alpha)$ for the corresponding factor.

Substituting $u_{\alpha\beta}(k)$ and $\varepsilon_{\alpha\beta}(k)$ in (5.77), $\delta H^\pm$ takes the form (9.2), where

$$H_1^\pm = \tfrac{1}{2} H_n u_0 [(\omega_{ms}/\omega_q)(e_z q_\pm + e_\pm q_z) + (\omega_a/\omega_q)(e_z q_\pm - e_\pm q_z)] \, . \tag{9.11}$$

We introduce three mutually perpendicular unit polarization vectors $e_\lambda (\lambda = 1, 2, 3)$; let $e_1 \| q$ correspond to longitudinal polarization, and the vectors e_2 and e_3 for the transverse polarization lie in the (XY) and (Zq) planes, respectively.*

We then obtain for the absorption coefficient (9.9), using (9.11),

$$\alpha^\lambda = \frac{\pi \chi_n f(\Omega) H_n^2 \Omega^2 \Omega_\lambda^2(\theta_q)}{16 \rho c_s^3 \omega_q^2} \, , \tag{9.12}$$

---

* Note that, strictly speaking, the introduction of longitudinal and transverse waves is possible only for solids with isotropic elastic properties. Therefore the subsequent treatment of the angular dependence of the absorption coefficient is mainly intended as an illustration.

where

$$\Omega_1(\theta_q) = \omega_{ms}\sin 2\theta_q, \quad \Omega_2(\theta_q) = (\omega_{ms} - \omega_a)\cos\theta_q, \quad \Omega_3(\theta_q) = \omega_{ms}\cos 2\theta_q - \omega_a \tag{9.13}$$

and

$$\omega_q = \omega_e + \omega_E(\Omega/\Omega_s)^2 \simeq \omega_e.$$

At resonance $\Omega = \omega_n$ and $f(\Omega) = 1/\pi\Gamma_n = T_2/\pi$ (Γ_n is the half-width of the NMR line). Assuming also that $|\gamma_n H_n| \simeq \omega_n$ (the case of the nuclei of magnetic atoms), and using (9.8), we obtain

$$\alpha^\lambda \simeq \frac{I(I+1)\,\omega_n^4 \Omega_\lambda^2(\theta_q)\,\hbar^2 N_n}{48\Gamma_n \rho c_s^3 \kappa T \omega_e^2}. \tag{9.14}$$

For a very rough estimate we shall assume $\Omega_\lambda \sim \omega_e$ (which corresponds to $\omega_{ms} \sim \omega_e$ or $\omega_a \sim \omega_e$, according as the magnetostriction constant G or the magnetic anisotropy constant K_1 is larger).

Let us consider the Mn^{55} nucleus in the ferrite $MnFe_2O_4$. At low temperatures (Heeger et al., 1963), $\omega_n \simeq 4 \cdot 10^9\ sec^{-1}$ and $1/\Gamma_n \sim 10^{-7}$ sec. Assuming further $I = \frac{5}{2}$, $c_s = 4 \cdot 10^5$ cm/sec, $\rho = 5$ g/cm^3, and $N_n \sim 10^{22}$ cm^{-3}, we obtain $\alpha \sim 10^{-3} \times T^{-1}$ cm^{-1}. For the Fe^{57} nuclei in yttrium ferrite–garnet ($Y_3Fe_5O_{12}$), for which $\omega_n \simeq 5 \cdot 10^8$ sec^{-1}, $1/\Gamma_n \sim 10^{-3}$ sec (at $T \sim 4°K$), $I = \frac{1}{2}$, and $N_n \sim 10^{20}$ cm^{-3} (for the natural concentration of Fe^{57}), α is roughly two orders of magnitude less. Thus NMAR can only be observed at sufficiently low temperatures.

Besides the direct observation of NMAR by the resonant absorption of ultrasound near the NMR frequency, there is another possible method for the detection of acoustic excitation of nuclear spins. This method entails measuring the change in the intensity of the ordinary NMR in an ultrasound field. If ultrasound of sufficient power is delivered to the specimen at the NMR frequency, the saturation effect observed when

$$\gamma_n^2 H_1^2 T_1 T_2 \gtrsim 1 \tag{9.15}$$

will lead to acoustic heating of the nuclear spin system, and as a result, the observed NMR intensity will decrease. In many cases, it is simpler to detect NMAR by the second method than by the first (in particular, if the sample material has a small concentration of magnetic nuclei). It was exactly by the second method that NMAR was discovered in the antiferromagnet $KMnF_3$ (see below).

It is not difficult to estimate the strain amplitude u_0 corresponding to the saturation condition (9.15). By (9.11),

$$u_0^2 \sim \frac{4\omega_e^2 c_s^2}{\Omega_\lambda^2 \omega_n^2 (\gamma_n H_n)^2 T_1 T_2}.$$

Substituting this u_0^2 in (9.6), we can find the sound energy flux needed to achieve

saturation of the nuclear spin system:

$$c_s \bar{E} \sim \frac{2\omega_e^2 \rho c_s^2}{(\gamma_n H_n)^2 \, \Omega_\lambda^2 T_1 T_2}. \qquad (9.16)$$

For $MnFe_2O_4$ ($T_1 \sim 10^{-3}$ sec) and $Y_3Fe_5O_{12}$ ($T_1 \sim 10^{-2}$ sec), we have $c_s \bar{E} \sim$ 10 W/cm^2 and 10^{-2} W/cm^2, respectively.

We shall now briefly examine how the above expressions for α and $c_s \bar{E}$ change when two other mechanisms of interaction between sound and nuclear spins are applied.

If the dependence of these quantities on the direction of sound propagation and the polarization of sound is irrelevant, then for the modulation mechanism it is simply necessary to put $\Omega_\lambda \equiv \omega_q$ in the corresponding expressions (9.12) and (9.16), while the order of magnitude of H_n will be determined by the anisotropic part of the HFI:[*]

$$H_n \sim \sum_j |A_j^\alpha - A_j^\beta| \, S/\hbar\gamma_n, \qquad (9.17)$$

where A_j^z are the components of the HFI tensor in the principal axes (see (5.87) and (5.88)). Note that the spin-acoustic interaction associated with changes in the interatomic distance (which is proportional to the strain tensor $u_{\alpha\beta}$) can be appreciable only when the equilibrium electron magnetization M does not coincide with any of the principal axes of the tensor $A^{\alpha\beta}$. This limitation does not apply to the interaction associated with the rotation of the crystal axes (which is proportional to the rotation tensor $\varepsilon_{\alpha\beta}$).

Unlike the strain mechanism, the rotation mechanism modulates the interaction of the nuclear spin not only with other atoms but also with the magnetic moment of its own atom.

An estimate of the order of magnitude of the NMAR effects for the quadrupole interaction mechanism can be obtained if in expressions (9.12) and (9.16) we put again $\Omega_\lambda \equiv \omega_q$, setting $H_n \sim (\hbar\gamma_n)^{-1} e^2 Q/a^3$ or $H_n \sim \omega_Q/\gamma_n$ for the interaction due to $u_{\alpha\beta}$ or $\varepsilon_{\alpha\beta}$, respectively ($Q$ is the quadrupole moment of the nucleus and ω_Q is the quadrupole frequency (6.1), which determines the quadrupole splitting for nuclei with noncubic surroundings).

3. NMAR in Antiferromagnets

We see from a comparison of (5.87) and (5.88) that the expression for the coefficient α in an antiferromagnet of the EA type will differ from (9.12) only by the additional factor $(\gamma_e H_E/\omega_k)^2$. In this case $\omega_k^2 \approx \omega_e^2 \approx |\gamma_e| H_E \omega_a$ and we obtain instead of (9.12)

$$\alpha^\lambda \simeq \frac{nI(I+1)\omega_n^2 (\gamma_n H_n)^2 \Omega_\lambda^2(\theta_q) \hbar^2}{48 c_s^3 \Gamma_n M_1 \omega_a^2 \kappa T}, \qquad (9.18)$$

---

[*] When the dipole interaction is modulated, $H_n \sim \mu_B/a^3 \sim M_0$.

where the frequency $\Omega_\lambda(\theta_q)$ is again determined by (9.13), M_1 is the molecular mass, and n is the number of the relevant nuclei in a molecule (we have used the equality $N_n/\rho = nM_1$).

Let us apply this equation to the evaluation of α for F^{19} nuclei in MnF_2. Because of the external field, ω_n for nonmagnetic atoms may differ substantially from $\gamma_n H_n$.

In particular, when $H\|Z$, we obtain for the two (magnetically) nonequivalent positions of the F^{19} nuclei in MnF_2 two different frequencies $\omega_n^{(1,2)} \approx \gamma_n|H_n \pm H|$. When $H \neq 0$, only half of the nuclei will resonate at the given frequency, and therefore $n = 1$ (although there are two fluorine atoms in a molecule).

Since the magnetostriction constant G is not known for MnF_2, while the anisotropy field H_a is very large, only terms with $\omega_a = |\gamma_e| H_a$ are retained in expression (9.13) for Ω_λ. Thus ω_a drops out from (9.18). Assuming in accordance with experimental data $\gamma_n \hbar H_n \sim \hbar \omega_n \sim 10^{-18}$ erg, $I = \frac{1}{2}$, $\Gamma_n \sim 10^5$ sec^{-1} (Kaplan et al., 1966), $M_1 = 1.7 \cdot 10^{-22}$ g, and $c_s \approx 4 \cdot 10^5$ cm/sec, we obtain $\alpha \sim 10^{-3}$ T^{-1} (cm^{-1}).

α is roughly one order of magnitude larger for the Mn^{55} nuclei in MnF_2 for which $I = \frac{5}{2}$, Γ_n^{-1} is two orders of magnitude smaller, and ω_n two orders of magnitude larger than for F^{19}.

Thus in antiferromagnets of the EA type, the NMAR effects are in general of the same order of magnitude as in ferromagnets.

The sound flux for which saturation effects become appreciable in antiferromagnets of the EA type is again determined by an expression of the form (9.16), with ω_e replaced by ω_a. In particular, for the F^{19} nuclei in MnF_2, at $T \sim 1°K$, when $T_1 \sim 10^2$ sec (Kaplan et al., 1966), and for the same values of the other parameters as before, we have $c_s \bar{E} \sim 10^{-4}$ W/cm^2.

As for spin-lattice relaxation, the NMAR effects associated with the magnetoelastic mechanism are especially large for antiferromagnets of the EP type which have a low-frequency spin wave branch.

Using relations (5.89), (9.2), and (9.10), we can write for these antiferromagnets the linearly polarized components of the circular amplitudes $H_1^\pm = H_1^\xi \pm iH_1^\eta$ (Figure 23),

$$H_1^\xi \equiv H_1^z = \frac{1}{2} u_0 H_n \frac{\gamma_e H_E \left[\omega_{ms}^{(2)}(e_y q_z + e_z q_y) + \omega_a(e_y q_z - e_z q_y)\right]}{\omega_{2q}^2}, \qquad (9.19)$$

$$H_1^\eta = \frac{1}{2} u_0 H_n^\xi \frac{\gamma_e H_E \omega_{ms}^{(1)}(e_y q_x + e_x q_y)}{\omega_{1q}^2}. \qquad (9.20)$$

Here $\omega_{ms}^{(1)}$ and $\omega_{ms}^{(2)}$ are two characteristic magnetostrictive frequencies, defined by the relations*

$$\omega_{ms}^{(1)} = \gamma_e G_1/M_0$$

---

* It is assumed that the external field H lies in the "easy plane" and its magnitude is such that it overcomes the weak anisotropy in this plane. The antiferromagnetic vector L is thus always perpendicular to the di-

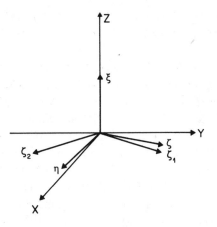

Figure 23

The coordinate axes ξ, η, and ζ associated with the nuclear spin quantization axis (ζ).
ζ_1 and ζ_2 are the directions of the equilibrium sublattice magnetizations.

and

$$\omega_{ms}^{(2)} = \gamma_e G_2/M_0 \,.$$

H_n^{ζ} is the projection of the hyperfine field $H_n = -\sum_j A_j \langle S_j\rangle/\hbar\gamma_n$ on the direction ζ of the nuclear spin quantization axis; ξ and η are unit vectors perpendicular to the ζ axis (one of which coincides with the Z axis).

For nuclei of magnetic atoms, when usually $|H_n| \gg H$, ζ virtually coincides with the direction of the respective magnetic sublattice magnetization (ζ_1 or ζ_2); accordingly $H_n^{\zeta} \simeq H_n = -A_0 M_0$. Since, on the other hand, ζ_1 and ζ_2 are respectively close to the positive and the negative direction of the antiferromagnetic axis $L \| Y$, the vector η is close to $H \| X$ and, consequently, $H_1^{\eta} \simeq H_1^{x}$.

Let us find the coefficient α for the two components of the field H_1, H_1^{ξ} (9.19) and H_1^{η} (9.20).

α is expected to have its largest value for H_1^{η}, since this component is associated with the interaction of the ultrasound with nuclear spins via the low-frequency spin wave branch (whose minimal frequency ω_{e1} can be determined for various

rection of H, which is taken as the X axis. Expressions (9.19) and (9.20) do not take into account the anisotropic part of magnetostriction, which depends on the direction of H in the basal plane. If the anisotropic part is taken into account, additional terms appear in (9.19) and (9.20). For example, for rhombohedral crystals of the $MnCO_3$ or α-Fe_2O_3 type, it is necessary to add in (9.19) and (9.20) terms of the form

$$\omega_{ms}^{(3)} \left[(e_z q_x + e_x q_z) \cos 3\psi - (e_z q_y + e_y q_z) \sin 3\psi \right],$$

$$\omega_{ms}^{(4)} \left[(e_x q_x - e_y q_y) \cos 3\psi - (e_x q_y + e_y q_x) \sin 3\psi \right],$$

respectively, where ψ is an angle defining the direction of H in the basal plane.

particular antiferromagnets of the EP type from (2.79), (2.83) and (2.86)). In this case

$$\alpha^\lambda = \frac{\pi\chi_n f(\Omega)(H_n^\zeta)^2 \omega_n^2 (\omega_{ms}^{(1)})^2 (\gamma_e H_E)^2 P_\lambda(\Theta_q, \Phi_q)}{16\rho c_s^3 \omega_{1q}^4}, \tag{9.21}$$

where

$$P_1(\Theta_q, \Phi_q) = \sin^2\Theta_q \sin^2 2\Phi_q,$$

$$P_2(\Theta_q, \Phi_q) = \sin^2\Theta_q \cos^2 2\Phi_q,$$

$$P_3(\Theta_q, \Phi_q) = \tfrac{1}{4}\sin^2 2\Theta_q \sin^2 2\Phi_q$$

(Θ_q and Φ_q are the polar and the azimuthal angles of the sound wave vector q).

Thus the optimum value of α is produced by ultrasound with the wave vector q and the polarization e lying in the "easy plane"; accordingly, for transverse waves q should be directed along H or L, and for longitudinal waves—at an angle of 45° to these vectors.*

For the low-frequency spin wave branch in antiferromagnets of the EP type, $|\gamma_e| H_E/\omega_{e1} \geqslant 10^2$ and therefore the NMAR effects, for the same values of the parameters, are four or more orders of magnitude larger than in antiferromagnets of the EA type or in ferromagnets.

For a rough estimate, we may take for Mn^{55} nuclei in compounds of the type $AMnF_3$ or $MnCO_3$,

$$\omega_{ms}^{(1)} \sim \omega_{1q} \sim \omega_{e1}, |\gamma_e| H_E/\omega_{e1} \sim 10^2, \Gamma_n \sim 10^7 \sec^{-1},$$

$$M_1 = N_n/\rho \sim 3\cdot 10^{-22}\,\text{g} \quad \text{and} \quad c_s \sim 4\cdot 10^5\,\text{cm/sec}.$$

As a result, we obtain $\alpha \sim 10/T$ (cm^{-1}).

The flux $c_s\bar{E}$ leading to the saturation of the nuclear spin-system, is given in this case, according to (9.6), (9.15) and (9.20), by the relation

$$c_s\bar{E} \sim \frac{2\rho\omega_{e1}^4 c_s^3}{(\omega_{ms}^{(1)})^2 (\gamma_e H_E)^2 (\gamma_n H_n)^2 T_1 T_2}. \tag{9.22}$$

There are unfortunately no experimental data on T_1 for these antiferromagnets. If we assume that the same mechanism of indirect interaction between nuclear spins and thermal lattice vibrations via spin waves is responsible for the spin-lattice relaxation, and substitute $1/T_1$ from (5.96) in (9.22), we obtain instead of (9.22),

$$c_s\bar{E} \sim \frac{\kappa T \omega_n^2}{3\pi T_2 c_s^2}.$$

---

* In the presence of magnetostriction dependent on the direction of H in the basal plane, an additional term which depends on the direction of H and has a different angular dependence on Θ_q and Φ_q (see footnote on p. 161) appears in (9.21).

This expression gives an extremely small value for the sound flux, $c_s \bar{E} \sim 10^{-9} T$ (W/cm^2).

Ultrasound experiments should also be carried out for Fe^{57} nuclei in hematite (α-Fe$_2$O$_3$), where the magnetostriction constant is comparatively large, so that $\omega_{ms}^{(1)} \simeq 2 \cdot 10^{12}$ sec^{-1} (see Chapter V, Sec. 7). Using the values of Matsuura et al. (1962) for $T_2 \sim 10^{-3}$ sec and $T_1 \sim 10^{-1}$ sec at room temperatures, where hematite is an antiferromagnet of the EP type with a weak ferromagnetism, we obtain

$$\alpha \sim 0.1 \text{ cm}^{-1} \quad \text{and} \quad c_s \bar{E} \sim 10^{-7} \text{ W/cm}^2 .$$

So far, we have not considered the correlation effects caused by the interaction between the nuclear spins via spin waves. For antiferromagnets of the EP type, these effects are particularly large at low temperatures. When they are taken into account, the $\chi(q, \Omega)$ in (9.4) is identified with the susceptibility of the nuclear spin wave (which can be calculated from equations of the form (2.52)). The situation is analogous, as was shown in Chapter III, to adding terms dependent on the gradients of M_1 and M_2 to the energy (2.13) and assuming that the variable magnetic field has the form $h = h_0 \exp(i\omega t - ikr)$. In addition, the correlation of the nuclear spins of course also affects the form of the effective fields H_1^ξ and H_1^η.

Note, however, that the component H_1^η (which is considerably larger than H_1^ξ for antiferromagnets of the EP type) for nuclei of magnetic atoms, when η is approximately parallel to H (Figure 23), will excite a second branch of nuclear spin vibration with the frequency Ω_{n2}. We have mentioned in Chapters II and III that the dynamic shift of the NMR frequency is small for this branch, since it interacts with the high-frequency spin wave branch (this does not apply to isotropic antiferromagnets of the RbMnF$_3$ or KMnF$_3$ type, for which both are low-frequency branches).

Also note that the ordinary NMR at the frequency Ω_{n2} is difficult to observe because there is no r-f field amplification effect for this vibration mode. Yet the resonant acoustic absorption must be quite large at this very frequency.

The antiferromagnetic resonance frequency shift due to hyperfine coupling apparently offers an additional opportunity for NMAR observations. Indeed, if a sound flux is delivered to the sample at the frequency Ω_{n2} and antiferromagnetic resonance is simultaneously observed at the frequency Ω_{e1}, we should register the disappearance of the shift when the flux reaches the value (9.22).

We shall not write the expression for the coefficient α corresponding to the component (9.19) $H_1^\xi \equiv H_1^z$ of the effective field. If $\omega_{e2} \gg \omega_{e1}$, as it is for uniaxial crystals of the MnCO$_3$, α-Fe$_2$O$_3$, or CsMnF$_3$ type, the NMAR effects here will be considerably smaller than for $H_1^\eta \approx H_1^x$.*

Since $\omega_{e2}^2 \approx |\gamma_e| H_E \omega_a$, α will have the form (9.18), but with a different angular

---

* However, if magnetostriction is very small, so that $\gamma_e H_E \omega_{ms}^{(1)} \ll \omega_{e1}^2$, it can be shown that the second term in (9.19), associated with rotation of the crystal axes, will predominate.

dependence $\Omega_\lambda(\Theta_q, \Phi_q)$. In such weakly anisotropic antiferromagnets as RbMnF$_3$ and KMnF$_3$, for which $\omega_{e2} \sim \omega_{e1}$, the contributions from both components of H_1 will be of the same order of magnitude.

Finally, the two other mechanisms of nuclear spin-acoustic coupling (modulation of the hyperfine and the dipole interaction, or the quadrupole mechanism) give for antiferromagnets (both EA and EP type) the same results as for ferromagnets. All that was stated in Sec. 2 in connection with these mechanisms could be repeated here almost in its entirety.

NMAR effects so far have been experimentally observed only at F^{19} nuclei in the antiferromagnet KMnF$_3$ (by means of acoustic quenching of the ordinary NMR). Denison et al. (1964) assume that the modulation mechanism (the part proportional to $u_{\alpha\beta}$) is responsible for the effects that they observed. However, additional experiments are needed to finally establish the nature of the phenomenon.

CHAPTER X

NMR Measurements in Magnetic Crystals

The instrumentation for NMR measurements in magnetic crystals (both magnetically ordered and paramagnetic) must comply with certain requirements connected with the specific features of the phenomenon in these crystals.

1. NMR Spectrometers for Paramagnets

The NMR line width in paramagnets may range from tens of oersteds for nuclei of nonmagnetic ions to hundreds of oersteds for nuclei of such ions as Co^{2+} (Shulman, 1959; Petrov and Nedlin, 1967) and Mn^{2+} (Jones, 1965). The corresponding relaxation parameters $1/T_1$ and $1/T_2$ can attain values of the order of 10^6 sec^{-1}. Additional line broadening occurs in polycrystals if the local field at the nucleus has an anistropic component. Because of the local field, the line may be strongly shifted (~ 100–1000 Oe) relative to its position for the corresponding nuclei in diamagnetic compounds.

For the observation of NMR signals in paramagnets, the same apparatus is used in principle as for the investigation of nonmagnetic solids (Lösche, 1963), i.e., the resonance is observed in an external field which is modulated and slowly scanned. Here no high stability and magnetic field homogeneity are required (stability and homogeneity of 10^{-5}–10^{-6} are quite satisfactory), but for paramagnets a somewhat wider range of magnetic field scanning and modulation amplitude is required. The basic problem is to ensure a sufficiently high sensitivity of the setup since the signals are very weak because of the large line width.

Since the spin-lattice relaxation time in paramagnets is sufficiently short, substantial gain in signal intensity can be obtained if the amplitude of the r-f field in the sample coil is increased. The signal-to-noise ratio (for the voltage) is proportional

to the amplitude of the field h_0. At the same time, the largest value of h_0 which yet does not lead to excessive saturation of the signal is determined by the condition $\gamma_n^2 T_1 T_2 h_0^2 \sim 1$. It is thus clear that for the NMR of F^{19} in MnF_2, where $T_1 = T_2 \approx 10^{-6}$ sec, the optimal field is $h_0 \approx 30$ Oe and the voltage across the sample coil may reach tens of volts (Shulman and Jaccarino, 1957). Therefore, for measurements in paramagnets, it is advisable to use autodynes, passive circuits (bridge or inductive) with a high voltage level, and superregenerative detectors. Pulse methods (spin echo) are very complicated to use in paramagnets because of the short relaxation times.

The sensitivity as measured by the signal-to-noise ratio can practically be the same for passive circuits and autodynes. The latter, however, are considerably simpler to build and operate. Besides, they are more readily tunable to different frequencies when the measurements are to be carried out on different nuclei. The advantage of passive circuits is that an optimal voltage attaining tens of volts can be applied to the sample without any deterioration in the sensitivity of the recording system, while for autodynes the sensitivity at such voltage amplitudes on the circuit drops because of nonlinearity in the tube transconductance. However, this advantage cannot always be put to work, because at such high amplitudes it is difficult to obtain the necessary stability for bridge balancing or noise compensation in the method of crossed coils.

Figure 24 shows a very convenient and fairly sensitive autodyne used in NMR measurements of F^{19}, $Tl^{203,\,205}$, Rb^{87}, Na^{23} and Co^{59} in paramagnets in the frequency interval from 6 to 35 MHz (Petrov, 1966). Earlier, Pound and Knight used a circuit based on analogous ideas. Frequency tuning was achieved by changing the inductance coil and adjusting the tuning capacitor C_1. When the frequency change was considerable, the positive feedback circuit also had to be tuned.

For work at high r-f field levels, the circuits suggested by Knight (1961) and Nagasawa (1964) can also be used.

Good results are obtained with superregenerative detectors, since they have a high sensitivity at large r-f field amplitudes and, as autodynes, are easy to tune in a wide frequency interval. The description of a superregenerative detector applied to NMR measurements on protons in $FeCl_2 \cdot 2H_2O$ (Narath, 1963) is given by Narath et al. (1964).

Superregenerative detectors are useful when weak, wide NMR lines are sought (as in MnO, where the Mn^{55} NMR line width is about 300 Oe), but they are inadequate for accurately measuring the position and the shape of the line. In a superregenerative detector, the high-frequency oscillations are periodically quenched. Quenching can be caused either by an external voltage or by the action of an e.m.f. originating in the oscillator. Therefore, the frequency spectrum generated by the superregenerative detector contains frequencies which are separated from the fundamental frequency by an amount ω_q (ω_q is the quench frequency which, in practice, constitutes tens or hundreds of kHz) and an NMR signal appears at each of the superregeneration

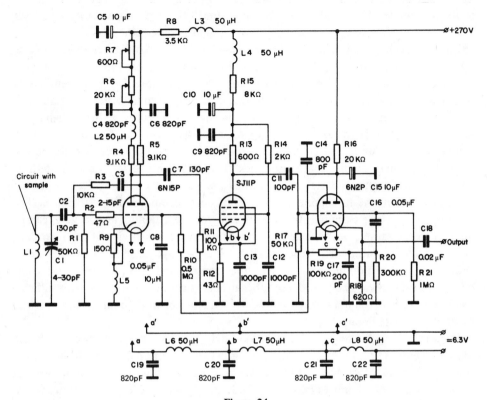

Figure 24

An autodyne circuit for NMR measurements of F^{19}, $Tl^{203,205}$, Rb^{87}, Na^{23} and Co^{59} in paramagnets in the frequency interval 6–35 MHz (Petrov, 1966).

frequencies. If the intensity of the signals is approximately the same, it is difficult to determine the resonance corresponding to the central frequency. Moreover a distortion of the line shape occurs.

2. Measurements in Magnetically Ordered Crystals

At temperatures below the magnetic ordering point the conditions for the observation of NMR sharply change. When the local fields at the nuclei are small, the spectrometers described in the previous section can be used. If the resonance is observed under conditions such that the internal local field is considerably larger than the external field, or when the external field has but a weak influence on the NMR frequency (for example, in a ferromagnet with a domain structure), the methods become essentially different.

First, the resonance can be observed without any external magnetic fields.

Second, the spectrometer should cover a wide frequency range, since the internal local magnetic field and, consequently, the NMR frequency changes from its maximal value to zero as the temperature is varied. The resonance frequency intervals for various nuclei in ferro- and antiferromagnets are approximately the following:

$$Fe^{57} \quad 40\ MHz-80\ MHz,$$

$$Mn^{55} \quad 100\ MHz-700\ MHz,$$

$$Co^{59} \quad 100\ MHz-300\ MHz,$$

$$Cr^{53} \quad 30\ MHz-\ 70\ MHz,$$

$$Ni^{61} \quad 20\ MHz-\ 40\ MHz.$$

For nuclei of the rare-earth metals, the resonance frequencies may lie in the centimeter range (Herve and Veillet, 1961; Itoh et al., 1968; Kobayashi et al., 1966, 1967a, b). For more detailed data on resonance frequencies see Appendix II.

Third, if continuous and not pulse methods of observation are used, frequency modulation and scanning of the resonance region are generally required.

Note that the apparatus for nuclear quadrupole resonance usually satisfies similar requirements (see, for example, Grechishkin and Soifer, 1964). Some problems related to the conditions of observation of NMR in local fields are discussed by Gossard et al. 1961a).

For NMR observations in local magnetic fields, both continuous and pulse methods are used. The choice of the particular method is determined by the actual experimental conditions, i.e., by the amplification coefficient η of the r-f field at the nucleus (see Chapter I, Sec. 2).

Continuous Wave Methods

In this case oscillator circuits (autodynes and superregenerative detectors) are most frequently employed, while in the microwave region slotted lines are used. Bridge circuits are hardly used as it is difficult to tune them in a wide frequency range.

Slotted lines are employed when the signal intensity is large enough. Heeger et al. (1964) and Houston and Heeger (1966) used a shorted coaxial line in which a few grams of the sample material were placed ($MnFe_2O_4$ or Mn_3O_4). The magnetic resonance of the Mn^{55} nuclei was observed. The reflected power was measured, whose magnitude changed on passing through the resonance. To increase the sensitivity, the fundamental power level was compensated by means of another line. The signal in $MnFe_2O_4$ was so strong that it could be easily observed on an oscillograph. Autodynes and superregenerative detectors have a high sensitivity and are suitable for the observation of very weak signals.

A Hopkins type autodyne circuit is often used (Buyle–Bodin, 1959). At frequencies below 70–80 MHz, oscillators based on Robinson's circuit (1959) are frequently used. The advantage of this circuit is in its stable operation at a low generation level. This is important for those materials in which the spin-lattice relaxation time is large and the amplification coefficient η is high. To avoid saturation, a low voltage should be applied to the sample coil.

Between 20 and 200 MHz a circuit similar to the push-pull marginal oscillation circuit (Benedek and Kushida, 1960; Shulman, 1961) is employed.

Measurements at still higher frequencies (100–700 MHz) can be carried out with oscillators using lengths of lines (La Force, 1961; Jeffers and Jones, 1965).

When searching for unknown signals in a wide frequency range, from units to hundreds MHz, it is convenient to use superregenerative detectors. They possess a high sensitivity: for example, O'Sullivan et al. (1963) observed NMR at the chlorine nuclei in $CuCl_2 \cdot 2H_2O$ by means of a superregenerative detector but could not observe it with an autodyne.

The optimal conditions for observation with superregenerative detectors depend on the amplitude of the variable field and also on the quench frequency and the nuclear spin relaxation time.

As has already been mentioned, the distinctive feature of measurements in local fields is the use of frequency modulation and scanning of the resonance region. Frequency modulation in autodynes is accomplished by including a variable capacitance in the circuit containing the sample (the capacitance varies with the modulation frequency). The modulation frequency is usually tens or hundreds of Hz.

A vibrating condenser, a reactive electron tube, or the capacitance of the space charge layer in a semiconductor diode may be used as the "oscillating" capacitance (Berman, 1963; Dutcher and Scott, 1961; Kudimov and Svetlov, 1966). The use of diodes is the more convenient since the frequency modulation unit becomes very compact, as is essential for a r-f circuit, and it is easy to tune the modulation frequency practically between any limits. For frequency modulation in a wide range, a varactor layer diode, silicon stabilitrons, and other diodes are employed. A circuit with a diode is shown in Figure 25. The d.c. bias is applied across the impedance Z_1, and the

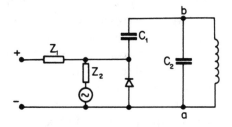

Figure 25

A circuit with a varactor diode.

modulating signal is applied to Z_2. Ohmic resistances (of the order of hundreds of kohm) or high frequency rectifier valves may be taken as Z_1 and Z_2. A capacitance C_1 of a few pF isolates the diode from the d.c. circuit and reduces the capacitance introduced into it by the diode. The frequency deviation in such a circuit is determined by the ratio $\frac{1}{2}\Delta C_{ab}/C_k$, where ΔC_{ab} is the change in capacitance across the terminals ab when a modulating voltage is applied to the diode and C_k is the total capacitance of the circuit. With such a circuit it is easy to attain frequency deviations of a few kHz or tens of kHz.

Frequency scanning of the resonance region can be accomplished by rotating the condenser by means of a motor or by varying the back voltage on the varactors.

One of the main shortcomings of autodynes with frequency modulation is the existence of a parasitic amplitude modulation accompanying the frequency modulation. The source of the parasitic amplitude modulation can be the frequency dependence of the feedback circuits, the Q-factor, and losses in the test specimen. Since the NMR signal is very weak, parasitic amplitude modulation, even if it constitutes a fraction of a percent of the generation level, may exceed the signal magnitude by a large factor. It is difficult to apply special devices to stabilize the level of generation since the useful signal modulates the autodyne oscillations exactly at that frequency at which the parasitic modulation is produced. Various methods are known for suppressing the parasitic amplitude modulation. We shall explain some of them.

Robinson (1963) suggested to make the frequency dependence of the feedback coefficient opposite to that of the equivalent resistance of the circuit. The generation level will thereby remain constant in a wide frequency range. In some cases the suppression of parasitic modulation is accomplished either by introducing an additional modulating voltage in counterphase with the parasitic modulation, or by applying a compensating voltage of appropriate phase and amplitude at the first amplifier stage (Dutcher and Scott, 1961; Volpicelli et al., 1965). It is, however, difficult to maintain this compensation with tunable autodynes, when the specimen is changed or the temperature is varied.

The most effective method of suppressing parasitic amplitude modulation is to measure the second or third harmonics of the signal. The intensity ratios A_1, A_2 and A_3 for the first, second and third harmonics are determined respectively by the following expressions (Grechishkin and Soifer, 1964):

$$A_1 = v_m \frac{df}{dv} + \left(\frac{v_m}{2}\right)^3 \frac{d^3f}{dv^3} + \ldots, \qquad A_2 = \frac{1}{4}v_m^2 \frac{d^2f}{dv^2} + \frac{1}{3}\left(\frac{v_m}{2}\right)^4 \frac{d^4f}{dv^4} + \ldots,$$

$$A_3 = \frac{1}{3}\left(\frac{v_m}{2}\right)^3 \frac{d^3f}{dv^3} + \frac{1}{12}\left(\frac{v_m}{2}\right)^5 \frac{d^5f}{dv^5} + \ldots$$

Here $f(v)$ is the absorption line shape function and v_m is the amplitude of the frequency deviation. Despite the considerable reduction in the intensity of the signal for small v_m this method is highly convenient for observation, since it does not require retuning of the compensating voltage when the frequency of the oscillator is substantially varied, the specimen is replaced, etc. To avoid overloading the low-frequency amplifier by the parasitic modulation signal, it is filtered out at the first amplifier stages.

The other elements of the measuring device, in particular the phase detector, are standard equipment as used in various r-f spectrometers.

Pulse Methods

In a number of cases, especially when the broadening of the NMR lines is highly nonuniform, and when the amplification coefficient η is very high, it is advisable to use pulse methods and, in particular, spin echo. Today the spin-echo method is frequently used for measurements in ferrites and ferromagnets.

The main advantages of pulse methods are the following:

1. Possibility of direct measurement of the relaxation times (T_1 and T_2).
2. Simple measurement of the amplification coefficient η.
3. Convenient searching for unknown lines, since r-f pulse of length τ_u covers a frequency interval $1/\tau_u$.

The spin-echo signal in nonuniformly broadened lines, moreover, is stronger than the NMR signal observable by continuous wave methods. Another important advantage is that in pulse methods simpler storage devices can be used.

In solids with short nuclear relaxation times (shorter than a millisecond) it is usually difficult to apply pulse methods, since short, powerful r-f pulses are required, together with a short recovery time for the receiver sensitivity. For example, to observe the maximal echo signal for Fe^{57} nuclei (by the Hahn method), the following conditions must be fulfilled:

$$\tau_u \ll T_2$$

and

$$\gamma_n h_0 \tau_u = 2\pi/3 . \tag{10.1}$$

If $\tau_u = 1\ \mu\text{sec}$ and $\gamma_n/2\pi = 137.7\ \text{Hz/Oe}$ (Ludwig and Woodbury, 1960), $h_0 \simeq 2.4\ \text{kOe}$ is required. To establish a field of such magnitude in the sample coil, the variable voltage in the circuit must be of the order of tens of kilovolts. In magnetic crystals, $h_0 = \eta h_{rf}$, where $\eta \sim 10^2$–10^4 and h_{rf} is the field set up in the sample coil. Therefore conditions for the creation of the required field h_0 at the nuclei are considerably simplified. If the magnitude of h_0 is known, it is easy to determine η from experiment (using (10.1)) by measuring the pulse length τ_u corresponding to the strongest echo

signal. A detailed description of a spin-echo spectrometer, operating at frequencies to 620 MHz, can be found in the work of Yasuoka (1964).

The Double Resonance Method

A block diagram of the double resonance system is shown in Figure 26. A ferromagnetic or antiferromagnetic resonance spectrometer is used.* A small sample coil (one turn or a few turns) is placed in a resonator in which microwave oscillations are excited. Voltage at the NMR frequency is fed into the coil from a separate oscillator. The coil should be oriented in such a way that the variable field h_{rf} of NMR frequency is perpendicular to the longitudinal component of

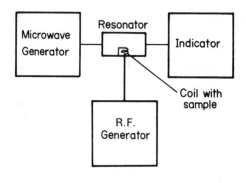

Figure 26

A block diagram of a double resonance system.

the hyperfine field at the nucleus. The ferromagnetic or antiferromagnetic resonance signal is recorded. When a voltage of NMR frequency is applied to the coil, the temperature of the spin-system increases and the nuclear longitudinal magnetization m_0 decreases. This leads, according to (2.58) and (2.80), to a shift in the electron resonance frequency and to an observable change in its intensity. The shift and the intensity of the observed signal depend on the frequency and the amplitude of the applied field h_{rf}. If h_{rf} is amplitude-modulated, the observed signal is also modulated and detection can be carried out using the modulation-frequency alternating current. To increase the sensitivity, conventional narrow-band devices, such as a phase detector, can be used.

Double resonance often permits measuring the nuclear spin-lattice relaxation time (Heeger and Houston, 1964). The change in the nuclear magnetization Δm when a modulating field h_{rf} of frequency Ω is applied is given by the equation

---

* See Heeger et al. (1961); Lee et al. (1963).

$$\Delta m_\Omega = \frac{\Delta m_0}{(1 + \Omega^2 T_1^2)^{1/2}} \tag{10.2}$$

Therefore, if the intensity of the double-resonance signal is measured as a function of the modulation frequency Ω, it is easy to determine the relaxation time T_1. Heeger et al. (1964) thus measured T_1 for the Mn^{55} nuclei in the ferrite $MnFe_2O_4$.

3. Increasing the Sensitivity of NMR Spectrometers

There are two ways for increasing the sensitivity of r-f spectrometers.

1. By reducing the bandwidth of the recording device through increasing the time constant of the synchronous (phase) detector.

2. By using storage devices which sum the results of repeated measurements (the time averaging method) (Klein and Barton, 1963; Jardetzky et al., 1963).

Both methods inevitably prolong the duration of the experiment; however, the second method has some advantages. Increasing the time constant of the synchronous detector increases the recording time for a single scanning of the resonance line. It is therefore necessary to ensure very high stability and low noise in all the recording elements at infralow frequencies. And yet the spectra of numerous noises (such as microphonics, contact noise in semiconductors, thermal drift, etc.) have a maximum at infralow frequencies. A synchronous detector cuts off the low-frequency noises of the preceding stages, but does not average the low-frequency modulation of the signal by the noise, and also introduces its own noise. Thus, when the time constant of the synchronous detector is increased, low-frequency noise becomes highly significant, and its spectral intensity increases when the noise frequency goes to zero.

In the second method, the line is scanned repeatedly at a fairly high speed and the results of all measurements are added. A wider passband is required for scanning, but it does not include frequencies close to zero. Noise with $\omega_{ns} \to 0$ (which is recorded all the same) shifts the signal line as one whole, without distorting its shape. The magnitude of the signal is proportional to the number of scans n, while the noise (assuming a "white" noise spectrum) is proportional to $\sqrt{n}$. For an arbitrary noise spectrum, with correlations taken into account, the dependence of noise on n may differ from the $\sqrt{n}$ law (Ernst, 1965). In the general case, for "white" noise, the application of the second method does not lead to an increase in the signal/noise ratio, if we neglect the slight gain in the signal obtained as a result of the optimal scanning speed (Ernst and Anderson, 1965). If the noise intensity increases when the frequency is lowered, time averaging gives a much higher sensitivity. In practice, multichannel analyzers perform time averaging (Reichert and Pereli, 1966; Zelenin et al., 1966). The channel switching sequence is synchronized with the magnetic field or frequency sweep. Noise and signal voltage control the number of pulses entering each channel. The device operates like a time analyzer.

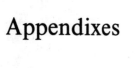

Appendixes

APPENDIX I
TABLE OF MAGNETIC NUCLEI

Nucleus	Spin	Magnetic moment (nuclear magnetons)	Quadrupole moment, 10^{-24} cm^2	Relative abundance %
H^1	1/2	2.7928	0	99.985
B^{11}	3/2	2.68858	0.36	80.39
F^{19}	1/2	2.6287	0	100
Al^{27}	5/2	3.64145	0.15	100
P^{31}	1/2	1.13166	0	100
Cl^{35}	3/2	0.82185	−0.079	75.53
Cl^{37}	3/2	0.68411	0.63	24.47
K^{39}	3/2	0.39140	0.09	93.10
K^{41}	3/2	0.21484	±0.11	6.88
Sc^{45}	7/2	4.7563	−0.22	100
Ti^{47}	5/2	−0.7884	—	7.32
Ti^{49}	7/2	−1.1036	—	5.46
V^{51}	7/2	5.1482	0.27	99.75
Cr^{53}	3/2	−0.47434	0.03	9.55
Mn^{55}	5/2	3.4678	0.32	100
Fe^{57}	1/2	0.0905	0	2.19
Co^{59}	7/2	4.6488	0.4	100
Ni^{61}	3/2	±0.75	—	1.25
Cu^{63}	3/2	2.2259	0.16	69.1
Cu^{65}	3/2	2.3852	0.19	30.9
Zn^{67}	5/2	0.87552	0.18	4.11
Ga^{69}	3/2	2.01603	0.19	60.4
Ga^{71}	3/2	2.56162	0.12	39.6
Ge^{73}	9/2	−0.87884	−0.2	7.76
As^{75}	3/2	1.439	0.3	100
Br^{79}	3/2	2.1057	0.33	50.537
Br^{81}	3/2	2.2695	0.28	49.463
Rb^{85}	5/2	1.35276	0.286	72.15
Y^{89}	1/2	−0.137316	0	100
Nb^{93}	9/2	6.1671	−0.22	100
Mo^{95}	5/2	−0.91334	—	15.7
Mo^{97}	5/2	−0.93252	—	9.45
Pd^{105}	5/2	−0.6015	—	22.33

APPENDIX I (continued)

Nucleus	Spin	Magnetic moment (nuclear magnetons)	Quadrupole moment, 10^{-24} cm^2	Relative abundance, %
Ag^{107}	1/2	−0.113548	0	51.35
Ag^{108}	1/2	−0.13054	0	48.65
Cd^{111}	1/2	−0.59501	0	12.75
In^{113}	9/2	5.5238	1.14	4.23
Sn^{117}	1/2	−0.99984	0	7.57
Sn^{119}	1/2	−1.04621	0	8.58
Sb^{121}	5/2	3.3590	−0.29	57.25
Sb^{123}	7/2	2.5466	0.68	42.75
Te^{125}	1/2	−0.88715	—	6.99
I^{127}	5/2	2.8090	0.78	100
Ba^{135}	3/2	0.83721	0.25	6.59
Ba^{137}	3/2	0.93657	0.2	11.32
La^{139}	7/2	2.7781	0.23	99.911
Pr^{141}	5/2	5.1	−0.054	100
Nd^{143}	7/2	−1.24	0.0206	12.17
Nd^{145}	7/2	0.77	0.0105	8.30
Sm^{147}	7/2	−0.76−1	0.23	15.07
Sm^{149}	7/2	−0.85	0.8	13.82
Eu^{151}	5/2	0.419	0.95	47.82
Eu^{153}	5/2	1.501	2.4	52.18
Gd^{155}	3/2	0.242	1.1	14.73
Gd^{157}	3/2	0.3225	1.0	15.68
Tb^{159}	3/2	1.52	—	100
Dy^{161}	5/2	−0.46	1.1	18.88
Dy^{163}	5/2	0.65	1.3	24.97
Ho^{165}	7/2	4.1	3.0	100
Er^{167}	7/2	0.58	2.8	22.94
Tm^{169}	1/2	−0.229	0	100
Yb^{171}	1/2	0.4930	0	14.31
Yb^{173}	5/2	−0.6725	3.9	16.13
Lu^{175}	7/2	2.236	5.68	97.41
Hf^{177}	7/2	0.5	3	18.56
Hf^{179}	9/2	−0.47	3	13.75
Ta^{181}	7/2	2.35	3.9	99.99
W^{183}	1/2	0.117216	0	14.4
Au^{197}	3/2	0.1449	0.58	100
Hg^{199}	1/2	0.502702	0	16.84
Hg^{201}	3/2	−0.556701	0.5	13.22
Tl^{203}	1/2	1.5960	0	29.50
Tl^{205}	—	1.62754	—	70.50
Pb^{207}	1/2	0.58954	0	22.6
Bi^{200}	9/2	4.0802	−0.34	100

APPENDIX II

EXPERIMENTAL DATA ON NMR IN FERRO- AND ANTIFERROMAGNETS
(RESONANCE FREQUENCIES AND LOCAL FIELDS)

Nucleus	Matrix	Temperature, °K	Frequency, MHz	Field	Notes and references
H^1	$CuCl_2 \cdot 2H_2O$ antiferro	~0	3.5	0.76	Paulis et al. (1958)
	$FeCl_2 \cdot 2H_2O$ antiferro	~0	2.10	2.168	Narath (1965a)
B^{11}	MnB ferro	300	32.4	23.7	Unpublished data E. Hirahara; Review (1965b)
F^{19}	MnF_2 antiferro	~0	160	40	Jaccarino and Shulman (1957)
	NiF_2 weakly ferro	4.2	100	10.075	Shulman (1961)
P^{31}	MnP ferro	80		74.0 / 69.7	Unpublished data E. Hirahara; Review (1965b)
Cl^{35}	$CuCl_2 \cdot 2H_2O$ antiferro	~0	10 / 19 / 28	28 / 28 / 28	Quadrupole triplet; O'Sullivan et al. (1963)
Cl^{35} / Cl^{37}	$CoCl_2 \cdot 2H_2O$ antiferro	4.0	16.415 / 13.460	27.558 / 27.595	Narath (1965b)
Cl^{35} / Cl^{37}	$FeCl_2$ antiferro	4.2		77	Jones and Segel (1964)
Se^{77}	$CdCr_2Se_4$ ferro	4.2	87.2 – 63.5		Berger et al. (1968) / Stauss et al. (1968)
	$HgCr_2Se_4$ ferro	4.2	81.8 – 58		Berger et al. (1968)
V^{51}	Fe ferro	273	94.7	84.6	Lam et al. (1963)

APPENDIX II (continued)

Nucleus	Matrix	Temperature, °K	Frequency, MHz	Field	Notes and references
Cr[53]	CrCl$_3$ ferro	~0	63.246	262.87	Layer structure, weakly antiferromagn. coupling between layers. Narath (1961)
	CrBr$_3$ ferro	~0	58.096 / 56.9	241.46 / 237	Center of triplet with intervals 295 kHz; Gossard et al. (1962)
	CrI$_3$ ferro	2	49.7 / 48.5	205.11 / —	From nuclei in domains; Narath (1965)
	CrO$_2$ ferro	~0	36.65	152.3	From nuclei in domain walls; Yasuoka et al. (1963)
	Cr$_2$O$_3$ antiferro	1.6	70.43	292.7	Rubinstein et al. (1964)
	MnCr$_2$O$_4$ noncollinear ferrite	~0	66.8 / 65.0	— / —	Two lines for the two types of sites; Nagasawa et al. (1965)
	CoCr$_2$S$_4$ ferro	77	47.4		Le Dang Khoi (1968)
	CuCr$_2$S$_4$ ferro	77	38.9		Le Dang Khoi (1968)
	CdCrS$_4$ ferro	4.2	43–47.4	−191	Berger et al. (1968) The isotropic part of the field is presented
	CdCr$_2$Se$_4$ ferro	4.2	40.8–45.4	−182	Berger et al. (1968) Stauss et al (1968) The isotropic part of the field is presented
	HgCr$_2$Se$_4$ ferro	4.2	40–44.4	−178	Berger et al. (1968) The isotropic part of the field is presented
Mn[55]	MnB ferro	300	217.7 / 203.6	206 / 193 }	Unpublished data, E. Hirahara; Review (1965b)
	Mn$_4$N ferro	282	117.9	111.7	Hihara et al. (1963)
	MnP ferro	82	143.4 / 116.4	136 / 110.8 }	Unpublished data, E. Hirahara; Review (1965b)

APPENDIX II (continued)

Nucleus	Matrix	Temperature, °K	Frequency, MHz	Field	Notes and references
Mn^{55}	Ni ferro	298	309	293	Koi and Tsujimura (1963)
	MnAs ferro	77	235.08	222.7	Lam et al. (1963)
	Mn_5Ge_3 ferro	90	201.8	192 }	Quadrupole splitting Hihara et al. (1963); Jackson et al. (1963)
	Mn_2Sn ferro	1.6	196–216	— }	
		82	250	237	Unpublished data; Review (1965b)
	MnSb	273	234.37	222	Tsujimura et al. (1962)
	Mn_2Sb ferro	82	126.26	119.6 }	Unpublished data, E. Hirahara; Review (1965b)
			143.7	137.1 }	
	MnBi ferro	77	234.4	222.1 }	Unpublished data, E. Hirahara; Review (1965b); Weinberg and Bloemberger (1960)
		196	225.0	213.3 }	
	$MnFe_2O_4$ ferro	4.2	585.1	554.5	Heeger et al. (1963), Yasuoka (1964)
		300	525		
	MnF_2 antiferro	~0	680	650	Jones and Jeffers (1964)
	$MnCr_2O_4$ ferrite	4.2	545.8		Houston and Heeger (1964)
	$Li_{0.75}\,Fe_{1.75}\,Mn_{0.75}\,O_4$ ferrite	90	310	293.7	Mn^{4+} ions in B-sites. Kubo et al. (1966)
	α-Mn antiferro	1.5	5	—	Itoh et al. (1963)
	Fe ferro	~0	239	227	Koi et al. (1964)
	Mn_3O_4 ferrite	4.2	557.4	—	Houston et al. (1966)
	MnO antiferro	1.5	617.8	—	Lines et al. (1965)

APPENDIX II (continued)

Nucleus	Matrix	Temperature, °K	Frequency, MHz	Field	Notes and references
Mn^{55}	KMnF$_3$ antiferro	4.2	676	339	Minkewicz and Nakamura (1966a)
					Frequency extrapolated to nonshifted value
	CsMnF$_3$ antiferro	4.2	666	—	Minkewicz and Nakamura (1966b)
					Frequency extrapolated to nonshifted values
Fe^{57}	Fe ferro	−0	46.65	339	Budnick et al. (1961); Robert and Winter (1960); Ogawa and Morimoto (1961)
		300	45.44	—	
	Co ferro	4.2	44.8	329	Budnick et al. (1963)
	α-Fe$_2$O$_3$ weakly ferro	273	71.5	525	Le Dang Khoi and Bertaut (1962); Matsuura et al. (1962)
		300	67.621	491	Tetrahedral sites; Ogawa and Morimoto (1962)
	Fe$_3$O$_4$ ferrite	300	63.549	462	Octahedral sites; Boyd and Slonczewskii (1962)
	Eu$_3$Fe$_5$O$_{12}$ ferrite-garnet	303	54.76	398	Tetrahedral sites
			67.84	493	Octahedral sites; Ogawa and Morimoto (1962)
	Gd$_3$Fe$_5$O$_{12}$ ferrite-garnet	20	65.74	—	Tetrahedral sites
			76.11	—	Octahedral sites; Buyle-Bodin et al. (1962)
		303	54.69	397	Tetrahedral sites; Ogawa and Morimoto (1962)
	Tb$_3$Fe$_5$O$_{12}$ ferrite-garnet	300	54.80	—	Tetrahedral sites; Ogawa and Morimoto (1962)
	Dy$_3$Fe$_5$O$_{12}$ ferrite-garnet	303	53.93	392	Tetrahedral sites
			67.29	489	Octahedral sites; Le Dang Khoi and Buyle-Bodin (1961)
	HO$_3$Fe$_5$O$_{12}$ ferrite-garnet	303	53.74	390	Tetrahedral sites
			67.15	488	Octahedral sites; Ogawa and Morimoto (1962)
	Er$_3$Fe$_5$O$_{12}$ ferrite-garnet	303	53.56	389	Tetrahedral sites
			66.94	486	Octahedral sites; Ogawa and Morimoto (1962)
	Tm$_3$Fe$_5$O$_{12}$ ferrite-garnet	20	64.75	—	Tetrahedral sites; Buyle-Bodin et al. (1962)
		303	53.51	387	Tetrahedral sites
			66.79	485	Octahedral sites; Ogawa and Morimoto (1962)

APPENDIX II (continued)

Nucleus	Matrix	Temperature, °K	Frequency, MHz	Field	Notes and references
Fe^{57}	Y$_3$Fe$_5$O$_{12}$ ferrite-garnet	20	64.90	—	Tetrahedral sites
		77	76.05	—	Octahedral sites; Buyle-Bodin et al. (1962)
			64.47	468	Tetrahedral
		300	75.71	550	Octahedral sites; Ogawa and Morimoto (1962)
			54.2	395	Tetrahedral sites
			67.4	485	Octahedral sites; Robert (1960)
	Sm$_3$Fe$_5$O$_{12}$	303	54.98	399	Tetrahedral sites
			67.97	494	Octahedral sites; Ogawa and Morimoto (1962)
	Yb$_3$Fe$_5$O$_{12}$ ferrite-garnet	303	52.97	385	Tetrahedral sites
			66.54	483	Octahedral sites; Ogawa and Morimoto (1962)
	Lu$_3$Fe$_5$O$_{12}$ ferrite-garnet	20	64.52	—	Tetrahedral sites
			75.65	—	Octahedral sites; Ogawa and Morimoto (1962)
	MnFe$_2$O$_4$ ferrite	296	54.18	395	Abe et al. (1961)
			58.78	429	
	NiFe$_2$O$_4$ ferrite	300	67.5	490	Tetrahedral sites
			72.0	525	Octahedral sites; Abe et al. 1963
	Li$_{0.5}$Fe$_{2.5}$O$_4$ ferrite	293	68.94	501	Tetrahedral sites; Ogawa et al. (1962)
			70.77	514	Octahedral sites; Yasuoka et al. (1962)
		298	68.89	492	Tetrahedral sites
			70.80	506	Octahedral sites; Le Dang Khoi and Bertaut (1962)
	Y$_3$Ca$_x$Fe$_{5-x}$O$_{12}$ ferrite garnet	4.77	50–80	—	Two lines for tetrahedral and octahedal sites as a function of x and T; Streever and Uriano (1965)
	Ba$_3$Fe$_{12}$O$_{19}$ ferrite	4.2	70	508	Streever (1968)
			72.5	526	The frequencies correspond to nonequivalent positions of the nuclei
			76	553	
			59	429	
	(1-x)Y$_3$Fe$_5$O$_{12}$ + xCa$_3$Fe$_{3.5}$V$_{1.5}$O$_{12}$ ferrite	77	63.3—64	460—467	Paugurt and Petrov (1968) Tetrahedral sites

APPENDIX II (continued)

Nucleus	Matrix	Temperature, °K	Frequency, MHz	Field	Notes and references
Fe^{57}	Bi$_{0.5}$Ca$_{2.5}$Fe$_{3.75}$V$_{1.25}$O$_{12}$	77	70.4 67.2 63.6		Petrov et al. Unpublished data, octahedral and tetrahedral sites
Co^{57}	Fe ferro	~0	289.2	286.3	La Force et al. (1962) Koi et al. (1961); La Force et al. (1961); Simanek and Sroubek (1962)
	Co f.c.	~0	217.2	215.0	Portis and Gossard (1960)
	Co f.c.o.	~0	228	226	La Force et al. (1963); Koi et al. (1962); Hardy (1961)
	Ni ferro	295	111.6	110.5	La Force et al. (1961 and 1962); Bennet and Streever (1962)
	Co—Pd ferro	4.2	215.6	213.4	Ehara and Tumono (1962)
	CoF$_2$ antiferro	~0	180	180	Jaccarino (1959)
Ni^{61}	Fe ferro	77	89.1	235	Streever et al. (1963)
	Co ferro	77	71.7	189	Bennet and Streever (1962); Streever et al. (1963)
	Ni ferro	~0	28.35	74.8	Bruner et al. (1961); Streever and Bennet (1963)
Cu^{63}	Fe ferro	295	26.02	69	Koi et al. (1961)
		273	240.0	212.7	Koi et al. (1962); Kushida et al. (1962)
	Co ferro	282	190.7	157.5	Koi et al. (1962); Kushida et al. (1962)
	Ni ferro	1.4	~52	~46	Asayama et al. (1963)
	CuCr$_2$S$_4$ ferro	77	100.6		Le Dang Khoi (1968)

APPENDIX II (continued)

Nucleus	Matrix	Temperature, °K	Frequency, MHz	Field	Notes and references
Cu^{65}	Fe ferro	273	257.1	212.7	Kushida et al. (1962)
	Co ferro	282	190.7	157.5	Kushida et al. (1962)
	Ni ferro	290		47.2	Asayama et al. (1963)
	$CuCr_2S_4$ ferro	77	107.8	47.2	Le Dang Khoi (1968)
Ga^{71}	$Y_3Ga_xFe_{5-x}O_{12}$ ferrite-garnet	4.77	25–38	—	Depends on x and T, Streever and Uriano (1965)
As^{75}	MnAs	77	206.35	283.2	Hihara et al. (1962)
Br^{79}	$CrBr_3$ ferro	1.48	53.774	—	Quadrupole splitting; Gossard et al. (1964)
			130.04	—	
Br^{81}			55.321	—	
			121.61	—	
Sn^{117}	MnSn ferro	77	196.9	124.8	Unpublished data, E. Hirahara; Review (1965b)
Sn^{119}	Mn_3Sn_2 ferro	77	322	200	Sato and Hoshiko (1964)
			324	200	Sato and Hoshiko (1964)
Sb^{121}	MnSb ferro	273	358.2	352.5	Tsujimura et al. (1962)
	Mn_2Sb ferro	196	230.7	228	Unpublished data, E. Hirahara; Review (1965b)
	MnNiSb ferro	297	281.28	276.7	Unpublished data, H. Suzuki and E. Hirahara; Review (1965b)
Sb^{123}	MnSb ferro	273	194.80	352.5	Tsujimura et al. (1962)

APPENDIX II (continued)

Nucleus	Matrix	Temperature, °K	Frequency, MHz	Field	Notes and references
Sb^{123}	Mn$_2$Sb ferro	196	125.7	228	Unpublished data, E. Hirahara; Review (1965b)
	MnNiSb ferro	297	152.70	276.7	Unpublished data, H. Suzuki and E. Hirahara; Review (1965b)
Eu^{151}	EuS ferro	−0	151.6	342	Charap and Boyd (1964)
Eu^{153}			343.0	342	
Gd^{155}	GdN ferro	2.2	44.204	370	Boyd and Gambino (1964)
Gd^{157}			58.700	370	
Tb^{159}	Tb ferro	77	3047	4000±1000	Herve and Veillet (1961)
	Tb ferro	1.5	3776 3108 2439		Kobayashi et al. (1967b) The splitting of the spectrum is caused by quadrupole interaction
	Gd ferro	1.5	3068		Kobayashi et al. (1967a)
Dy^{161}	Dy ferro	1.4	1603 1219 830 445	5900±10%	Kobayashi et al. (1966) The splitting of the spectrum is caused by quadrupole interaction
	Gd ferro	1.5	817		Kobayashi et al. (1967a)
Dy^{163}	Dy ferro	1.4	1984 1574 1163 755 347	5900±10%	Kobayashi et al. (1966) The splitting of the spectrue is caused by quadrupole interaction
	Gd ferro	1.5	1144		Kobayashi et al. (1967a)

APPENDIX II (continued)

Nucleus	Matrix	Temperature, °K	Frequency, MHz	Field	Notes and references
Nd^{143}	Gd ferro	1.5	834		Kobayashi et al. (1967b)
Nd^{145}	Gd ferro	1.5	519		Kobayashi et al. (1967b)
Sm^{147}	Gd ferro	1.5	624		Kobayashi et al. (1967b)
Sm^{149}	Gd ferro	1.5	516		Kobayashi et al. (1967b)
Er^{167}	Gd ferro	1.5	896		Kobayashi et al. (1967b)
Tm^{169}	Gd ferro	1.5	2223		Kobayashi et al. (1967b)

Bibliography

ABE, H., M. MATSUURA, H. YASUOKA, A. HIRAI, T. HASHI, and T. FUKUYAMA (1963) *J. Phys. Soc. Japan*, **18**, 1400.

ABE, H. (1965) *J. Phys. Soc. Japan*, **20**, 267.

ABKOWITZ, M. and I.J. LOWE (1966) *Phys. Rev.*, **142** (2), 333.

ABRAGAM, A. (1963 [Russian transl.]) *The Principles of Nuclear Magnetism.* Oxford Univ. Press, 1961.

AKHIEZER, A.I., V.G. BAR'YAKHTAR, and M.I. KAGANOV (1960). *UFN*, **71**, 533; **72**, 3.

AKHIEZER, A.I., V.G. BAR'YAKHTAR, and S.V. PELETMINSKII (1967) *Spinovye volny (Spin Waves).* Moscow, "Nauka."

ALEXANDROV, I.V. (1964) *Teoriya yadernogo magnitnogo rezonansa (The Theory of Nuclear Magnetic Resonance).* Moscow, "Nauka."

ANDERSON, P.W. (1952) *Phys. Rev.*, **86**, 694.

ARGYRES, P.N. and P.L. KELLEY (1964) *Phys. Rev.*, **134**, A98.

ASAYAMA, K., SH. KOBAYASHI, and J. ITOH (1963) *J. Phys. Soc. Japan*, **18**, 458.

ASAYAMA, K. (1963) *J. Phys. Soc. Japan*, **18**, 1739.

BAKER, J.M. and W. HAYES (1957) *Phys. Rev.*, **106**, 603.

BALLHAUSEN, C.J. (1964 [Russian transl.]) *Introduction to the Ligand Field Theory.* New York, McGraw-Hill, 1962.

BAR'YAKHTAR, V.G., M.A. SAVCHENKO, V.V. GANN, and P.V. RYABKO (1964) *ZhETF*, **47**, 1968.

BAR'YAKHTAR, V.G., S.V. PELETMINSKII, and E.G. PETROV (1968) *FTT*, **10**, 785.

BELOV, K.P. (1959) *Magnitnye prevrashcheniya (Magnetic Transformations).* Moscow, Fizmatgiz.

BELOV, K.P. and I.S. LYUBUTIN (1965) *ZhETF* (letters), **1**, 26; **49**, 747.

BENEDEK, G.B. and J. AMSTRONG (1961) *J. Appl. Phys., Suppl.*, **32**, 106.

BENEDEK, G.B. and T. KUSHIDA (1960) *Phys. Rev.*, **118**, 46.

BENNET, L.H. (1965)   *Phil. Mag.*, **12**, 213.

BENNET, L.H. (1966)   *J. Appl. Phys.*, **37**, 1242.

BENNET, L.H. and R.L. STREEVER (1962)   *J. Appl. Phys., Suppl.*, **33**, 1093.

BENOIT, H. and J.P. RENARD (1964)   *Phys. Lett.*, **8**, 32.

BERGER, S.B., J.I. BUDNICK, and T.J. BURCH (1958)   *J. Appl. Phys.*, **39**, 658.

BERMAN, L.S. (1963)   *Nelineinaya poluprovodnikovaya emkost' (Nonlinear Semiconductor Capacitance)*. Moscow, Fizmatgiz.

BLOCKER, T.G. and A.J. HEEGER (1966)   *J. Appl. Phys.*, **37**, 1236.

BLOCH, F. (1930)   *Z. Phys.*, **61**, 206.

BLYUMENFEL'D, L.A., V.V. VOEVODSKII, and A.G. SEMENOV (1962)   *Primenenie elektronnogo paramagnitnogo rezonansa v khimii (The Application of Electron Paramagnetic Resonance in Chemistry)*. Novosibirsk.

BOGOLYUBOV, N.N. and S.V. TYABLIKOV (1949)   *ZhETF*, **19**, 256.

BOROVIK–ROMANOV, A.S. (1962)   *Antiferromagnetizm i ferrity (Antiferromagnetism and Ferrites)*. In series: "*Itogi Nauki,*" Izd. AN SSSR.

BOROVIK–ROMANOV, A.S., M.M. KREINES, and L.A. PROZOROVA (1963)   *ZhETF*, **45**, 64.

BOROVIK–ROMANOV, A.S. and E.G. RUDASHEVSKII (1964)   *ZhETF*, **47**, 295.

BOROVIK–ROMANOV, A.S. and V.A. TULIN (1965)   *ZhETF* (letters), **1**, 18.

BOUTRON, F. and C. ROBERT (1961)   *Compt. rend.*, **253**, 433.

BOYD, E.L. (1963)   *Bull. Amer. Phys. Soc.*, **8**, 439.

BOYD, E.L. (1966)   *Phys. Rev.*, **145**. 174.

BOYD, E.L. and J. GAMBINO (1964)   *Phys. Rev. Lett.*, **12**, 20.

BOYD, E.L. and J.C. SLONCZEWSKI (1962)   *J. Appl. Phys., Suppl.*, **33**, 1077.

BUDNICK, J.I., L.J. BRUNER, R.J. BLUME, and E.L. BOYD (1961)   *J. Appl. Phys., Suppl.*, **32**, 120.

BUDNICK, J.I., R.C. LA FORCE, and E.F. DAY (1963)   *Proc. XI. Colloq. AMPERE (Eindhoven, 1962)*, p. 629.

BUDNICK, J.I. and S. SKALSI (1967)   In: *Hyperfine Interactions*. A.J. FREEMAN and R.B. FRANKEL, Ed. New York–London, Academic Press.

BUISHVILI, L.L. (1963 a)   *FTT*, **5**, 229.

BUISHVILI, L.L. (1963 b)   *FTT*, **5**, 1027.

BUISHVILI, L.L. (1863 c)   *FTT*, **5**, 3291.

BUISHVILI, L.L. (1964)   *FTT*, **6**, 903.

BUISHVILI, L.L. and N.P. GIORGADZE (1963)   *FTT*, **5**, 1814.

BUISHVILI, L.L. and N.P. GIORGADZE (1965)   *FTT*, **7**, 769.

BUISHVILI, L.L., N.P. GIORGADZE, and G.E. GURGENISHVILI (1964)   *FTT*, **6**, 2238.

BUISHVILI, L.L., L.B. VATOVA, and N.P. GIORGADZE (1966)   *FTT*, **8**, 1309.

BURGIEL, J.C., V. JACCARINO, and A.L. SCHOWLOW (1961)   *Phys. Rev.*, **122**, 429.

BUYLE–BODIN, M. (1959)   *J. phys. radium*, **20**, 159A.

BUYLE–BODIN, M. and LE DANG KHOI (1962)   *J. phys. radium*, **23**, 565.

CALLEN, H.B., D. HONE, and A. HEEGER (1965)   *Phys.Lett.*, **17**, 233.

CHARAP, S.H. and E.L. BOYD (1964)   *Phys. Rev.*, **133A**, 811.

CHIRKOV, V.K. and A.A. KOKIN (1960)   *ZhETF*, **39**, 1381.

190   **BIBLIOGRAPHY**

COLE, P.H. and W.J. INCE (1966)   *Phys. Rev.*, **150**, No. 2, 377.

COLLECTION (1951): *Fizika ferromagnitnykh oblastei (The Physics of the Ferromagnetic Phase)* —a collection of translations. S.V. VONSOVSKII, Ed. Moscow, Innostrannoi Literatury.

COLLECTION (1962): *Effekt Messbauera (The Mossbauer Effect)*—a collection of translated Papers. YU. M. KAGAN, Ed. Moscow, Innostrannoi Literatury.

Collection (1963): *Teoriya ferromagnetizma metallov i splavov (The Theory of Ferromagnetism of Metals and Alloys)*—a collection of papers translated from English. S.V. VONSOVSKII, Ed. Moscow, Innostrannoi Literatury.

COLLINS, M.F. and G.G. LOW (1965)   *Proc. Phys. Soc.*, **86**, 535.

COWANT, D.L. and L.W. ANDERSON (1965)   *Phys. Rev.*, **139**, A424.

DASH, J.G., B.D. DUNLAR, and D.G. HUNLAP (1966)   *Phys. Rev.*, **141**, 376.

DAVIS, H.L. and A. NARATH (1964)   *Phys. Rev.*, **134A**, 433.

DE GENNES, P.G. and F. HARTMAN-BOUTRON (1961)   *Compt. rend.*, **253**, 2922.

DE GENNES, P.G., P.A. PINCUS, F. HARTMAN-BOUTRON, and M. WINTER (1963)   *Phys. Rev.*, **129**, 1105.

DENISON, A.B., J.M. JAMES, J.D. CURRIN, W.H. TANTIELA, and R.J. MAHLER (1964)   *Phys. Rev. Lett.*, **12**, 244.

DUTCHER, C.H. and T.A. SCOTT (1961)   *Rev. Sci. Instr.*, **32**, 457.

DYSON, F. (1956)   *Phys. Rev.*, **102**, 1217, 1230.

DUTCHER, C.H. and T.A. SCOTT (1961)   *Rev. Sci. Instr.*, **32**, 357.

DZYALOSHINSKII, I.E. (1957)   *ZhETF*, **32**, 1547; **33**, 1454.

EHARA, SH. (1964)   *J. Phys. Soc. Japan*, **19**, 1313.

EHARA, SH. and Y. TUMONO (1962)   *J. Phys. Soc. Japan*, **17**, 726.

EIBSCHITZ, M., S. STRIKMAN, and D. TREVES (1966)   *Solid State Comm.*, **4**, 141.

ELWELL, D. (1964)   *Proc. Phys. Soc.*, **84**, 409.

ERNST, R.R. (1965)   *Rev. Scient. Instr.*, **36**, 1689.

ERNST, R.R. and P.W. ANDERSON (1965)   *Rev. Scient. Instr.*, **36**, 1695.

FARZTDINOV, M.M. (1966)   *FMM*, **21**, 487.

FERMI, E. and T. SEGRÈ (1933)   *Z. Phys.*, **82**, 729.

FINK, H. and D. SHALTIEL (1964)   *Phys. Rev.*, **136A**, 218.

FISHER, M.E. (1965)   *Proc. Intern. Conf. Magnetism (Nottingham, 1964).*

FREEMAN, A.J. and R.E. WATSON (1961)   *Phys. Rev. Lett.*, **6**, 343.

FREEMAN, A.J. and R.E. WATSON (1965)   *Magnetism*, Vol. IIA, G.T. RADO and H. SUHL, Ed., New York–London, Academic Press.

GABUDA, S.P., A.G. LUNDIN, YU. V. GAGARINSKII, L.R. BATSANOVA, and L.A. KHRIPIN (1966)   *ZhETF*, **51**, 707.

GAMMEL, J., W. MARSHALL, and L. MORGAN (1963)   *Proc. Roy. Soc.*, **A275**, 257.

GOLDANSKII, V.I., V.A. TRUKHTANOV, M.N. DEVISHEVA, and V.F. BELOV (1965)   *Phys. Lett.*, **15**, 317.

GONDAIRA, K.-I. (1966)   *J. Phys. Soc. Japan*, **21**, 933.

GOSSARD, A.C. and A.M. PORTIS (1959) *Phys. Rev. Lett.*, **3**, 164.

GOSSARD, A.C., A.M. PORTIS, and W.I. SANDLE (1961 a) *J. Phys. Chem. Solids*, **17**, 341.

GOSSARD, A.C., V. JACCARINO, and J.P. REMEIKA (1961 b) *Phys. Rev. Lett.*, **7**, 122.

GOSSARD, A.C., V. JACCARINO, and J.P. REMEIKA (1962) *J. Appl. Phys., Suppl.*, **33**, 1187.

GOSSARD, A.C., V. JACCARINO, E.D. JONES, J.P. REMEIKA, and R. SLUSHER (1964) *Phys. Rev.*, **135A**, 1051.

GRECHISHKIN, V.S. and G.B. SOIFER (1964) *PTE*, **I**, 5.

GUREVICH, A.G. (1960) *Ferrity na sverkhvysokikh chastotakh (Ferrites at Ultrahigh Frequencies)*. Moscow, Fizmatgiz.

HALL, T.P., W. HAYES, R.W. STEVENSON, and J. WILKENS (1963) *J. Chem. Phys.*, **39**, 35.

HARDY, W.A. (1961) *J. Appl. Phys., Suppl.*, **32**, 112.

HEEGER, A.J., A.M. PORTIS, D.T. TEANEY, and C.L. WITT (1961) *Phys. Rev. Lett.*, **7**, 308.

HEEGER, A.J., S.K. GHOSH, and T.D. BLOCKER (1963) *J. Appl. Phys.*, **34**, 1034.

HEEGER, A.J., T.D. BLOCKER, and S.K. GHOSH (1964) *J. Appl. Phys.*, **35**, No. 3 (II), 840; *Phys. Rev.*, **134A**, 399.

HEEGER, A.J. and D.T. TEANEY (1964) *J. Appl. Phys.*, **35**, 846.

HEEGER, A.J. and T.W. HOUSTON (1964) *J. Appl. Phys.*, **35**, 836.

HEEGER, A.J. and T.W. HOUSTON (1965) *Proc. Intern. Conf. Magnetism (Nottingham, 1964)*, p. 395.

HELLER, P. (1966) *Phys. Rev.*, **146**, 403.

HELLER, P. and G.B. BENEDEK (1962) *Phys. Rev. Lett.*, **8**, 428.

HELLER, P. and G.B. BENEDEK (1965 a) *Phys. Rev. Lett.*, **14**, 71.

HELLER, P. and G.B. BENEDEK (1965 b) *Proc. Intern. Conf. Magnetism (Nottingham, 1964)*, p. 97.

HERVÉ, J. and P. VEILLET (1961) *Compt. rend.*, **252**, 99.

HERVÉ, J. and J.N. AUBRUN (1962) *J. phys. radium*, **23**, 570.

HIHARA, T., A. TSUJIMURA, and Y. KOI (1962) *J. Phys. Soc. Japan*. **17**, 1320.

HIHARA, T., Y. KOI, and A. TSUJIMURA (1963) *J. Phys. Soc. Japan*, **18**, 454.

HOLSTEIN, T. and H. PRIMAKOFF (1940) *Phys. Rev.*, **58**, 1098.

HONE, D., H. CALLEN, and L.R. WALKER (1966) *Phys. Rev.*, **144**, 283.

HONMA, A. (1966) *Phys. Rev.*, **142**, 306.

HOUSTON, T.W. and A.J. HEEGER (1964) *Phys. Lett.*, **10**, 29.

HOUSTON, T.W. and A.J. HEEGER (1966) *J. Appl. Phys.*, **37**, 1234.

HUANG, N.L., R. ORBACH, and E. SIMÁNEK (1966) *Phys. Rev. Lett.*, **17**, 134.

HUBBARD, J., D.E. RIMMER, and F.R.A. HORGONN (1966) *Proc. Phys. Soc.*, **88**, 13.

IGNATCHENKO, V.A. and YU.A. KUDENKO (1966) *Izv. An SSSR*, physics series, **30**, 77, 933.

ITOH, J. and Y. MASUDA (1963) *J. Phys. Soc. Japan*, **18**, 455.

ITOH, J., K. ASAYAMA, and S. KOBAYASHI (1965) *Proc. Intern. Conf. Magnetism (Nottingham, 1964)*, p. 382.

ITOH, J., S. KOBAYASHI, and N. SENO (1968) *J. Appl. Phys.*, **39**, 1325.

IZYUMOV, YU.A. and M.V. MEDVEDEV (1966) *FTT*, **8**, 2117.

192   BIBLIOGRAPHY

JACCARINO, V. (1959)   *Phys. Rev. Lett.*, **2**, 163; *J. Chem. Phys.*, **30**, 1627.

JACCARINO, V. (1965 a)   *Magnetism*, Vol. II A, G.T. RADO and H. SUHL, Ed. New York–London, Academic Press.

JACCARINO, V. (1965 b)   *Proc. Intern. Conf. Magnetism, (Nottingham, 1964)*, p. 377.

JACCARINO, V. (1968)   *J. Appl. Phys.*, **39**, No. 2, part II, 1166.

JACCARINO, V. and R.G. SHULMAN (1957)   *Phys. Rev.*, **107**, 1196.

JACCARINO, V., B.G. MATTIAS, M. PETER, H. SUHL, and J.H. WERNICK (1960)   *Phys. Rev. Lett.*, **5**, 251.

JACCARINO, V., L.R. WALKER, and G.K. WERTHEIM (1964)   *Phys. Rev. Lett.*, **13**, 752.

JACCARINO, V., N. KAPLAN, R.E. WALSTEDT, and J.H. WERNICK (1966)   *Phys. Lett.*, **23**, No. 9, 514.

JACKSON, R.F., R.G. SCURLOCK, D.B. UTTON, T.H. WILMSHURST, and M. RUBINSTEIN (1963)   *Phys. Lett.*, **6**, 39.

JACKSON, R.F., R.G. SCURLOCK, D.B. UTTON, and T.H. WILMSHURST (1965)   *Proc. Intern. Conf. Magnetism, (Nottingham, 1964)*, p. 384.

JANAK, J.F. (1964)   *Phys. Rev.*, **134A**, 411.

JARDETZKY, O., N.G. WADE, and J.J. FISHER (1963)   *Nature*, **197**, 183.

JEFFERS, K.B. and E.D. JONES (1965)   *Rev. Scient. Instr.*, **36**, 983.

JONES, E.D. (1965 a)   *J. Appl. Phys.*, **36**, part 2, 919.

JONES, E.D. (1965 b)   *J. Phys. Soc. Japan*, **20**, 1292.

JONES, E.D. and K.B. JEFFERS (1964)   *Phys. Rev.*, **135A**, 1277.

JONES, W.N., and J.S.L. SEGEL (1964)   *Phys. Rev. Lett.*, **13**, 528.

KAMIMURA, H. (1966)   *J. Phys. Soc. Japan*, **21**, 484.

KAPLAN, N., P. PINCUS, and V. JACCARINO (1966)   *J. Appl. Phys.*, **37**, 1239.

KARUYOSKI, H. (1964)   *J. Phys. Soc. Japan*, **19**, 1678, 2357.

KEFFER, F., T. OGUCHI, W. O'SULLIVAN, and J. JAMASHITA (1959)   *Phys. Rev.*, **115**, 1553.

KITTEL, CH. (1963 [Russian transl.])   *Introduction to Solid-State Physics.* New York, J. Wiley, 1959 (2nd ed.).

KLEIN, M.P. and C.W. BARTON (1963)   *Rev. Scient. Instr.*, **34**, 754.

KNIGHT, W.D. (1961)   *Rev. Scient. Instr.*, **32**, 95.

KOBAYASHI, SH., and J. ITOH (1965)   *J. Phys. Soc. Japan*, **20**, 1741.

KOBAYASHI, SH., N. SANO, and J. ITOH (1966)   *J. Phys. Soc. Japan*, **21**, 1456.

KOBAYASHI, SH., N. SANO, and J. ITOH (1967 a)   *J. Phys. Soc. Japan*, **23**, 474.

KOBAYASHI, SH., N. SANO, and J. ITOH (1967 b)   *J. Phys. Soc. Japan*, **22**, 676.

KOI, Y., A. TSUJIMURA, T. HIHARA, and T. KUSHIDA (1961)   *J. Phys. Soc. Japan*, **16**, 574, 1040.

KOI, Y., A. TSUJIMURA, T. HIHARA, and T. KUSHIDA (1962)   *J. Phys. Soc. Japan, Suppl.*, **17**, B-1, 88.

KOI, Y. and A. TSUJIMURA (1963)   *J. Phys. Soc. Japan*, **18**, 1347.

KOI, Y., A. TSUJIMURA, and T. HIHARA (1964)   *J. Phys. Japan*, **19**, 1493.

KRAMERS, H.A. (1934)   *Physica*, **1**, 182.

KUBO, R. (1952)   *Phys. Rev.*, **87**, 568.

KUBO, T., H. YASUOKA, and A. HIRAI (1966)   *J. Phys. Soc. Japan*, **21**, 812.

KUDYMOV, G.G. and YU.G. SVETLOV (1966)   *Radiospektroskopiya*, **XI**, 83.

KURKIN, M.I. and N. PARFENOVA (1966) *FMM*, **8**, 1839.
KURKIN, M.I. and E.A. TUROV (1967) *FMM*, **24**, 27.
KUSHIDA, T., A.H. SILVER, Y. KOI, and A. TSUJIMURA (1962) *J. Appl. Phys., Suppl.*, **33**, No. 3, 1079.

LA FORCE, R.C. (1961) *Rev. Scient. Instr.*, **32**, 1386.
LA FORCE, R.C., S.F. RAVITZ, and G.F. DAY (1961) *Phys. Rev. Lett.*, **6**, 226.
LA FORCE, R.C., S.F. RAVITZ, and G.F. DAY (1962) *J. Phys. Soc. Japan, Suppl.*, **17**, B-I, 99.
LA FORCE, R.C., L.E. TOTH, and S.F. RAVITZ (1963) *J. Phys. Chem. Solids*, **24**, 729.
LAM, D.J., D.O. VAN OSTENBURG, M.V. NAVITT, H.D. TRAPP, and D.W. PRACHT (1963) *Phys. Rev.*, **131**, 1428.
LANDAU, L.D. (1937) *ZhETF*, **7**, 19.
LANDAU, L.D. and E.M. LIFSHITS (1935) *Sow. Phys.*, **8**, 153.
LANDAU, L.D. and E.M. LIFSHITS (1963) *Kvantovaya mekhanika (Quantum Mechanics).* Moscow, Fizmatgiz.
LANDAU, L.D. and E.M. LIFSHITS (1964) *Statisticheskaya fizika (Statistical Physics).* Moscow, Fizmatgiz.
LANDAU, L.D. and E.M. LIFSHITS (1965) *Teoriya uprugosti (Theory of Elasticity).* Moscow, FIZMATGIZ.
LE DANG KHOI (1965) *Compt. rend.*, **261**, 1807.
LE DANG KHOI (1966) *Compt. rend.*, **262**, 1166.
LE DANG KHOI and E. BERTAUT (1962) *Compt. rend.*, **254**, 1584.
LE DANG KHOI (1968) *Solid State Comm.*, **6**, 203.
LEE, K., A.M. PORTIS, and L.G. WITT (1963) *Phys. Rev.*, **132**, 144.
LINES, M.E. and E.D. JONES (1965) *Phys. Rev.*, **139**, 1313.
LÖSCHE, A. (1963 [Russian transl.]) *Kerninduktion.* Berlin, Deutscher Verlag d. Wissenschaften, 1957.
LOUNASMAA, O.V. (1967) In: *Hyperfine Interactions,* A.J. FREEMAN and R.B. FRANKEL, Ed., New York–London, Academic Press.
L'VOV, V.G. and M.P. PETROV (1966) *Phys. stat. solidi*, **13**, K 65.
LOW, W. (1962 [Russian transl.]) *Paramagnetic Resonance in Solids.* New York, Academic Press, 1960.
LOW, G.G. (1966) *Phys. Lett.*, **21**, 497.
LUDWIG, G.W. and H.H. WOODBURY (1960) *Phys. Rev.*, **117**, 1286.

MARSHALL, W. and R. STUART (1961) *Phys. Rev.*, **123**, 2048.
MATSUURA, M.M. (1966) *J. Phys. Soc. Japan*, **21**, 886.
MATSUURA, M., H. YASUOKA, A. HIRAI, and T. HASHI (1962) *J. Phys. Soc. Japan,* **17**, 1147.
MATTIS, D. (1967 [Russian transl.]) *The Theory of Magnetism.* New York, Harper and Row, 1965.
MAYS, I. (1963) *Phys. Rev.*, **131**, 38.
MINKEWICZ, V. and A. NAKAMURA (1966 a) *Phys. Rev.*, **143**, 356.
MINKEWICZ, V. and A. NAKAMURA (1966 b) *Phys. Rev.*, **143**, 361.
MISETICH, A.A. and R.E. WATSON (1966) *Phys. Rev.*, **143**, 335.

194   BIBLIOGRAPHY

MITCHEL, A.H. (1957)   *J. Chem. Phys.*, **27**, 17.

MIZOGUCHI, T. and M. INOUE (1966)   *J. Phys. Soc. Japan*, **21**, 1310; *Techn. Rep. ISSR*, **A**, No. 190, Tokyo.

MIZOGUCHI, T. and R.E. WATSON (1965)   *J. Appl. Phys.*, **36**, part 2, 1020.

MORI, H. and K. KAWASAKI (1962)   *Progr. Theoret. Phys.*, **27**, 592.

MORIYA, T. (1956)   *Progr. Theoret. Phys.*, **16**, 23, 641.

MORIYA, T. (1962)   *Progr. Theoret. Phys.*, **28**, 371.

MORIYA, T. (1963)   In: *Magnetism*, Vol. I, G.T. RADO and H. SUHL, Ed. New York, Academic Press.

MORIYA, T. (1964)   *J. Phys. Soc. Japan*, **19**, 681.

MULLIKEN, R.S. (1933)   *Phys. Rev.*, **43**, 279.

MURRAY, G. and W. MARSHALL (1965)   *Proc. Intern. Conf. Magnetism (Nottingham, 1964).*

NAGASAWA, H. (1964)   *Jap. J. Appl. Phys.*, **3**, 476.

NAGASAWA, H. and T. TSUSHIME (1965)   *Phys. Lett.*, **15**, 205.

NAKAMURA, A. (1958)   *Progr. Theor. Phys.*, **20**, 542.

NAKAMURA, A., V. MINKEWICZ, and A.M. PORTIS (1964)   *J. Appl. Phys.*, **35**, 842.

NARATH, A. (1961)   *Phys. Rev. Lett.*, **7**, 410.

NARATH, A. (1963)   *Phys. Rev.*, **131**, 1929.

NARATH, A. (1965 a)   *Phys. Rev.*, **139**, 1221.

NARATH, A. (1965 b)   *Phys. Rev.*, **140**, 552.

NARATH, A. (1965 c)   *Phys. Rev.*, **140A**, 854.

NARATH, A. (1967)   In: *Hyperfine Interactions*, A.J. FREEMAN and R.B. FRANKEL, Ed. New York–London, Academic Press.

NARATH, A., W.I. O'SULLIVAN, W.A. ROBINSON, and W.W. SIMMONS (1964)   *Rev. Scient. Instr.*, **35**, 476.

NARATH, A. and H.L. DAVIS (1965)   *Phys. Rev.*, **137A**, 163.

OGAWA, S. and S. MORIMOTO (1961)   *J. Phys. Soc. Japan*, **16**, 2065.

OGAWA, S. and S. MORIMOTO (1962)   *J. Phys. Soc. Japan*, **17**, 654.

OGAWA, S., S. MORIMOTO, and Y. KIMURA (1962)   *J. Phys. Soc. Japan*, **17**, 1671.

OGUCHI, T. (1960)   *Phys. Rev.*, **117**, 117.

ONOPRIENKO, L.G. (1964)   *FMM*, **18**, 678.

ONOPRIENKO, L.G. (1965)   *FMM*, **19**, 481.

ONSAGER, L. (1944)   *Phys. Rev.*, **65**, 117.

O'REILY, D.E. and T. TSANG (1964)   *J. Chem Phys.*, **40**, 734.

O'SULLIVAN, W.J., W.A. ROBINSON, and W.W. SIMMONS (1961)   *Phys. Rev.*, **124**, 1317.

O'SULLIVAN, W.J., W.W. SIMMONS, and W.A. ROBINSON (1963)   *Phys. Rev. Lett.*, **10**, 476.

PAKE, G.E. (1965 [Russian transl.])   *Paramagnetic Resonance*. New York, W.A. Benjamin, 1962.

PAUGURT, A.P., and M.P. PETROV (1968)   *FTT*, **10**, 3451.

PAUL, D.I. (1962)   *Phys. Rev.*, **126**, 78; **127**, 455.

PAYNE, R.E., R.E. FORMAN, and A. KAHN (1965)   *J. Chem. Phys.*, **42**, 3806.

PENNEY, W.G. (1938 [Russian transl.]) *The Quantum Theory of Valency*. London, Methuen, 1936; VAN VLECK, J.H. and A. SHERMAN, *Rev. Mod. Phys.*, **7**, 167 (1935 [included in Russian transl. of Penney's work]).

PETROV, M.P. (1965) *FTT*, **7**, 1663.

PETROV, M.P. (1966) Thesis, Institut poluprovodnikov AN SSSR, Leningrad.

PETROV, M.P. and V.A. KUDRYASHOV (1966) *FTT*, **8**, 3124.

PETROV, M.P. and V.V. MOSKALEV (1968) In: *Yadernyi magnitnyi rezonans. (Nuclear Magnetic Resonance)*, P.M. BORODIN, Ed. Leningrad, Izd. LGU. (2nd ed.).

PETROV, M.P. and G.M. NEDLIN (1967) *FTT*, **9**, 3246.

PETROV, M.P. and G.A. SMOLENSKII (1965) *FTT*, **7**, 2156.

PETROV, M.P., G.A. SMOLENSKII, and P.P. SYRNIKOV (1965) *FTT*, **7**, 3699.

PETROV, M.P., and G.A. SMOLENSKII (1966) *ZhETF*, **50**, 871.

PINCUS, P. (1963) *Phys. Rev.*, **131**, 1530.

PINCUS, P. and J. WINTER (1963) *Phys. Rev. Lett.*, **7**, 269.

PORTIS, A.M. and A.C. GOSSARD (1960) *J. Appl. Phys.*, **31**, 205.

PORTIS, A.M. and J. KANAMORI (1962) *J. Phys. Soc. Japan*, **17**, 587.

PORTIS, A.M., G.L. WITT, and A.J. HEEGER (1963) *J. Appl. Phys.*, **34**, 1052.

POULIS, N.J. and G.E. HARDEMANN (1952) *Physica*, **18**, 201, 315.

POULIS, N.J. and G.E. HARDEMANN (1953) *Physica*, **19**, 391.

POULIS, N.J., G.E. HARDEMANN, W. VAN DER LUGHT, and W.P. HAAS (1958) *Physica*, **24**, 280.

POUND, P.V. and W.D. KNIGHT (1950) *Rev. Scient. Instr.*, **21**, 219.

PU FU-CHO (1960) *DAN SSSR*, **130**, 1244; **131**, 546.

REICHERT, J.F. and J. PERELI (1966) *Rev. Sci. Instr.*, **37**, 426.

REVIEW (1962)—WINTER, G.M. *J. Phys. Rad.*, **23**, 556.

REVIEW (1963)—ANDERSON, P.W. *Solid State Physics*, Vol. 14, p. 99; *Magnetism*, Vol. I, G.T. RADO and H. SUHL, Ed. New York–London, Academic Press.

REVIEW (1965 a)—JACCARINO, V. *Magnetism*, Vol. II A, G. T. RADO and H. SUHL, Ed. New York–London, Academic Press.

REVIEW (1965 b)—PORTIS, A. M. and R. H. LINQUIST *Magnetism*, Vol. II A, G. T. RADO and H. SUHL, Ed., New York–London, Academic Press.

REVIEW (1967)—NARATH, A. *Hyperfine Interactions*, A.J. FREEMAN and R.B. FRANKEL, Ed., New York–London, Academic Press.

RIMMER, D.E. (1965) *Proc. Intern. Conf. Magnetism (Nottingham, 1964)*, p. 337.

ROBERT, C. (1960) *Compt. rend.*, **251**, 2684.

ROBERT, C. (1961) *Compt. rend.*, **252**, 1442.

ROBERT, C. (1962) Thesis: *"Contribution à l'étude de la résonance nucléaire dans les corps ferromagnétiques."* University of Paris.

ROBERT, C. and J.M. WINTER (1960) *Arch. sci. (Geneva)*, **13**, 433.

ROBERT, C. and J.M. WINTER (1961) *Compt. rend.*, **253**, 2925.

ROBERT, C. and F. HARTMAN–BOUTRON (1962) *J. phys. radium*, **23**, 574.

ROBINSON, F.N.H. (1959) *J. Scient. Instr.*, **36**, 481.

ROBINSON, F.N.H. (1963) *Rev. Scient. Instr.*, **34**, 1260.

RUBINSTEIN, M., G.H. STAUSS, and J.J. KREBS (1964) *Phys. Lett.*, **12**, 302.

RUDERMAN, M.A. and C. KITTEL (1954) *Phys. Rev.*, **96**, 99.

SATO, N. and R. HOSHIKO (1964) *J. Phys. Soc. Japan*, **19**, 139.
SAVATSKY, E. and M. BLOOM (1964) *Canad. J. Phys.*, **42**, 657.
SCHERMER, R.I. and L. PASSEL (1965) *Bull. Amer. Phys. Soc.*, **10**, 75.
SEAVEY, M.H. and P.E. TANNENWALD (1958) *Phys. Rev. Lett.*, **1**, 168.
SEIDEN, I. (1962) *Compt. rend.*, **254**, 234.
SENTURIA, S.D. and G.B. BENEDEK (1966) *Phys. Rev. Lett.*, **17**, 475.
SHALTIEL, D. (1966) *Phys. Rev.*, **142**, 300.
SHULMAN, R.G. (1959) *Phys. Rev. Lett.*, **2**, 459.
SHULMAN, R.G. (1961 a) *Phys. Rev.*, **121**, 125.
SHULMAN, R.G. (1961 B) *J. Appl. Phys., Suppl.*, **32**, 126.
SHULMAN, R.G. (1966) *Physics*, **2**, 217.
SHULMAN, R.G. and V. JACCARINO (1956) *Phys. Rev.*, **103**, 1126.
SHULMAN, R.G. and V. JACCARINO (1957) *Phys. Rev.*, **108**, 1219.
SHULMAN, R.G. and V. JACCARINO (1958) *Phys. Rev.*, **109**, 1084.
SHULMAN, R.G. and K. KNOX (1960 a) *Phys. Rev. Lett.*, **4**, 603.
SHULMAN, R.G. and K. KNOX (1960 b) *Phys. Rev.*, **119**, 94.
SHULMAN, R.G. and K. KNOX (1965) *J. Chem. Phys.*, **42**, 813.
SHULMAN, R.G. and J.M. STOUT (1959) *Phys. Rev.*, **118**, 1136.
SHULMAN, R.G. and S. SUGANO (1963) *Phys. Rev.*, **130**, 506.
SILVERSTEIN, S.D. (1963) *Phys. Rev.*, **132**, 997.
SIMANEK, E. (1963) *Czech. J. Phys.*, **B13**, 732.
SIMANEK, E. and Z. SROUBEK (1964) *Phys. stat. solidi*, **4**, 251.
SLICHTER, C. (1967 [Russian transl.]) *Principles of Magnetic Resonance*. New York, Harper and Row, 1963.
SMOLENSKII, G.A., M.P. PETROV, V.V. MOSKALEV, V.S. LVOV, V.S. KASPEROVICH, and E.V. ZHIRNOVA (1968) *FTT*, **10**, 1305.
SMOLENSKII, G.A., V.M. YUDIN, P.P. SYRNIKOV, and A.B. SHERMAN (1966) *FTT*, **8**, 2965.
SPENCE, R.D., P. MIDDENTS, Z. ELSAFAR, and R. KLEINBERG (1964) *J. Appl. Phys.*, **35**, 854.
STAUSS, G.H., M. RUBINSTEIN, J. FEINLEIB, K. DWIGHT, N. MENYUK, and A. WOLD (1968) *J. Appl. Phys.*, **39**, 667.
STREEVER, R.L. (1963) *Phys. Rev. Lett.*, **10**, 232.
STREEVER, R.L. and L.H. BENNET (1963) *Phys. Rev.*, **131**, 2000.
STREEVER, R.L., L.H. BENNET, R.C. LA FORCE, and G.F. DAY (1963) *J. Appl. Phys.*, **34**, 1050.
STREEVER, R.L. and G.A. URIANO (1964) *Phys. Rev. Lett.*, **12**, 614.
STREEVER, R.L. and G.A. URIANO (1965) *Phys. Rev.*, **139A**, 305.
STREEVER, R.L. (1968) *Phys. Lett.*, **27A**, 563.
STUART, R.N. and W. MARSCHALL (1966) *Proc. Phys. Soc.*, **87**, 749.
SUGANO, S. and Y. TANABE (1965) *J. Phys. Soc. Japan*, **20**, 1115.
SUGANO, S. and R.G. SHULMAN (1963) *Phys. Rev.*, **130**, 517.
SUHL, H. (1958) *Phys. Rev.*, **109**, 606.
SUHL, H. (1959) *J. phys. radium*, **20**, 333.

TIERSTEN, H.F. (1964)   *J. Math. Phys.*, **5**, 1298.

TORNLEY, J.H.M., C.G. WINDSOR, and J. OWEN (1965)   *Proc. Roy. Soc.*, **284**, 252.

TSANG, T. (1964)   *J. Chem. Phys.*, **40**, 729.

TSUJIMURA, A., T. HIHARA, and J. KOI (1962)   *J. Phys. Soc. Japan*, **17**, 1078.

TUROV, E.A. (1963)   *Fizicheskie svoistva magnitno-uporyadochennykh kristallov (Physical Properties of Magnetically Ordered Crystals)*. Moscow, Izd. AN SSSR.

TUROV, E.A. (1967 a)   *FTT*, **9**, 1562.

TUROV, E.A. and N.G. GUSEINOV (1960)   *ZhETF*, **38**, 1326.

TUROV, E.A. and V.G. KULEEV (1965)   *ZhETF*, **49**, 248.

TUROV, E.A., M.I. KURKIN, and O.B. SOKOLOV (1967)   *FMM*, **23**, 786.

TUROV, E.A. and M.I. KURKIN (1968)   *FTT*, **10**, 785.

TUROV, E.A. and V.E. NAISH (1960)   *FMM*, **9**, 10.

TUROV, E.A. and A.I. TIMOFEEV (1967)   *ZhETF, letters*, **5**, 133.

TUROV, E.A. and V.G. SHAVROV (1965)   *FTT*, **7**, 217.

TYABLIKOV, S.V. (1965)   *Metody kvantovoi teorii magnetizma (Methods of the Quantum Theory of Magnetism)*. Moscow, "Nauka."

URIANO, G.A. and R.L. STREEVER (1965)   *Phys. Lett.*, **17**, 205.

VALIEV, K.A. (1957)   *Uchenye Zapiski Kazansk. Gos. Univ.*, **117** (9), 149.

VAN KRANENDONK, J. and M. BLOOM (1956)   *Physica*, **22**, 545.

VAN LOEF, J.J. (1966)   *Solid State Comm.*, **4**, 625.

VAN DER LUGHT, W. and N.I. POULIS (1960)   *Proc. 7th Internat. Conf. Low Temperature (Toronto, 1960)*, Amsterdam.

VAN DER LUGHT, W., N.J. POULIS, T.W.J. AGT, and C.J. GORTER (1962)   *Physica*, **28**, 195.

VAN VLECK, J.H. (1935)   *J. Chem. Phys.*, **3**, 803, 807.

VAN VLECK, J.H. (1941)   *J. Chem. Phys.*, **9**, 85; *J. phys. radium*, **12**, 262.

VAN VLECK, J.H. (1948)   *Phys. Rev.*, **74**, 1168.

VATOVA, L.B. (1965)   *FTT*, **7**, 2133.

VLASOV, K.B. (1960)   *ZhETF*, **38**, 889.

VOLPICELLI, R.I., B.D. NAGESWARA, and I.D. BALDESCHWIELER (1965)   *Rev. Scient. Instr.*, **36**, 150.

VOLSTEDT and VERNIK (1966).

VONSOVSKII, S.V. (1953)   *Sovremennoe uchenie o magnetizme (Modern Theory of Magnetism)*. Moscow, GITTL.

WATSON, R.E. and A.J. FREEMAN (1961)   *Phys. Rev.*, **123**, 2027. [Russian transl. in "Collection", 1963].

WATSON, R.E. and A.J. FREEMAN (1964)   *Phys. Rev.*, **134A**, 1526.

WATSON, R.E. and A.J. FREEMAN (1967)   In: *Hyperfine Interactions*. A.J. FREEMAN and R.B. FRANKEL, Ed., New York–London, Academic Press.

WEGER, M., E.L. HAHN, and A.M. PORTIS (1961)   *J. Appl. Phys., Suppl.*, **32**, 124.

WEINBERG, D.L. and N. BLOEMBERGEN (1960)   *Phys. Chem. Solids*, **15**, 240.

WELSH, L.B. (1966)   *Phys. Rev. Lett.*, **17A**, 3.

198   BIBLIOGRAPHY

WELSH, L.B. (1967)   *Phys. Rev.*, **156**, 370.

WHITE, R.M. (1965)   *J. Appl. Phys.*, **36**, 3653.

WILSON, G.V.H. (1964 a)   *J. Sci. Instr.*, **41**, 436.

WILSON, G.V.H. (1964 b)   *Proc. Phys. Soc.*, **84**, 689.

WINTER, J.M. (1961)   *Phys. Rev.*, **124**, 452.

WITT, G.L. and A.M. PORTIS (1964 a)   *Phys. Rev.*, **135**, 1616.

WITT, G.L. and A.M. PORTIS (1964 b)   *Phys. Rev.*, **136A**, 1316.

WOLFRAM, T. and J. CALLAWAY (1963)   *Phys. Rev.*, **130**, 2207.

WOLFRAM, T. and W. HALL (1966)   *Phys. Rev.*, **143**, 284.

WOLKER, M.B. (1966)   *Proc. Phys. Soc.*, **87**, 45.

WOLKER, M.B. and R.W.H. STEVENSON (1966)   *Proc. Phys. Soc.*, **87**, 35.

YASUOKA, H. (1964)   *J. Phys. Soc. Japan,* **19**, 1182.

YASUOKA, H. (1966)   *J. Phys. Soc. Japan,* **21**, 393.

YASUOKA, H., A. HIRAI, M. MATSUURA, and T. HASHI (1962)   *J. Phys. Soc. Japan,* **17**, 1071.

YASUOKA, H., H. ABE, A. HIRAI, and T. HASHI (1963)   *J. Phys. Soc. Japan,* **18**, 593.

ZELENIN, V.P., T.G. KUDYMOV, YU.G. SVETLOV, and B.A. TROSHEV (1966)   *Radiospektro-skopiya*, **XI**, 71.

ZIMAN, J.M. (1966 [Russian transl.])   *Principles of the Theory of Solids.* Cambridge University Press, 1964.

Subject Index